GALILEO, SCIENCE AND THE CHURCH

GALILEO, SCIENCE AND THE CHURCH

by
Jerome J. Langford, O.P.

Foreword by Stillman Drake

Desclee Company
New York

≈≋≈ To My Parents ≈≋≈

Revisores Ordinis: James A. Wesiheipl, o.p.
Benedict Ashley, o.p.
Imprimi Potest: Gilbert J. Graham, o.p., Prior Provincialis
Nihil Obstat: Edward J. Montano, s.t.d., Censor Librorum
Imprimatur: ✠ Terence J. Cooke, v.g.
New York, February 3, 1966

Contents

FOREWORD

In society as a whole, there are no dead issues. When we think there are, it is only because each of us moves in a necessarily restricted circle and remains untouched by the labors of persons outside it on matters that are of no interest to us and our immediate associates. The American Civil War has recently been reopened in our literature to a degree that astounds those of us who thought it to be a dead issue. Doubtless those who avidly read that literature are equally astonished by the current revival of interest in the Galileo affair.

But as Father Langford points out in his opening sentence, the Catholic Church has not been allowed to regard its ancient condemnation of Galileo as a dead issue. The effective prohibition of a scientific thesis which later became an established truth has stood ever since as a symbol. It is a considerable part of Father Langford's task to unravel the various threads that have been woven into that symbol over the centuries. With this part of his task I have great sympathy, for it seems to me that many of the most formidable problems of contemporary society stem from unfortunate forms taken by symbolic modes of expression. Galileo's condemnation was the first of two outstanding events in modern times that have come to symbolize the conflict of religion and science. Indeed, without the treatment accorded to the views of Galileo and Darwin, the very concept of such a conflict might never have been formulated. Once conceived, however, the inevitability of such a conflict has grown to the status of a commonplace, despite many highly articulate protests in our own time by men eminent in either discipline.

Being neither a scientist nor a religious man, I have no direct way of knowing whether or not there is an inherent conflict between the two modes of thought. Galileo seems to be an example to the contrary; certainly a first-rate scientist, he regarded himself as a good Catholic, and the Church itself, in the very document by which it condemned him, acknowledged that he had behaved as one throughout his trial.

On the other hand, I do know that there is an inherent conflict between established authority and independent thought. The independent thinker is always condemned, one way or another, by established authority, which usually asserts that it does this not only for the sake of truth and the good of society, but also for his own good. The social patterns of the time may determine whether this benefit is conferred on its recipient by execution, imprisonment or merely social disapproval, but the fact of antagonism does not change. Now before the birth of modern science, there was no objective way to tell whether a particular independent thinker was a genius or a crank, and the safest course for established authority was to suppress him on general principles, leaving it to posterity to determine whether he had been a prophet or a madman. The procedure is not unknown today in the absence of objective criteria; that is, outside the truly scientific fields. That is why I think that if Galileo's case symbolizes anything, it symbolizes the inherent conflict between authority and freedom rather than any ineradicable hostility of religion toward science. It was an accident of Galileo's time that authority happened to be vested in a particular religious institution and that his field of independent thought happened to be the creation of modern science. If that is so, then I see no reason why that institution as it exists today, which I suppose to be rather different from what it was three or four centuries ago, should be singled out for its having put into practice in his case the only general principle by which unquestioned authority can be maintained by those so unfortunate as to possess it. No more should it be singled out for credit in having kept alive what science there was during the ten or fifteen centuries immediately preceding, though this is perhaps something that should be explained by those who believe in the inherent conflict of religion and science.

Of course it has always been irksome to independent thinkers, whether geniuses or cranks, to have limits placed by established authorities on their activities. Whether on the whole the imposition of such limits has been good for truth and for society is another question. Probably a man's opinion on that question is directly proportional to his abilities as an independent thinker, or at any rate to his own estimate of those abilities. What is too often overlooked is that most men are not independent thinkers, and would prefer not to be bothered by them. It is probably this that accounts for the fact that Galileo still has plenty of enemies today, a fact which used to puzzle me but for which I am very grateful, as otherwise I should have nothing to write about. Mr. Arthur Koestler is perhaps the best living representative of the pro-authority and anti-Galileo forces, though he is no spokesman for the Church; if I understand his writings correctly, he believes that the Church authorities were on the whole better able than Galileo to conduct the affairs of his time, and that we might do well to follow their example and suppress the intrusion of daydreaming scientists into public affairs in our own day. It is interesting to see that Father Langford, on the other hand, believes that the Church made a mistake in the case of Galileo, and reasonably asks that the circumstances be properly understood, so that the Church of today may be spared further attacks based on an error of such venerable age. One may hope that after another four centuries, the institutions which manage to survive their current authoritarian mistakes in the name of truth and society will in their turn ask our descendents not to judge them by the actions of their remote ancestors who suppressed independent thought and unproved theories.

Galileo, of course, did not attack the Church, but the Church felt that its authority was threatened by his views. Galileo thought the precise opposite, and went to much trouble to warn it of the probable consequences if it officially condemned a theory which might later be proved true. This aspect of Galileo's behavior in 1614–1616, when his warnings put him in some personal peril, is neglected by most writers, who read his warnings as demands. Thus Father Langford, like Mr. Koestler, sees the theme of Galileo's *Letter to Christina* as a shift in the burden of proof. To my mind, that document was exclusively a plea against the possible prohibition of the Copernican theory, and was by no means a plea for active support of that theory by the Church. Galileo was not a man to plead for support, though he was one to fight against interference. And he said quite clearly at the time that what he opposed was not the prohibition of Copernicus's book, but its prohibition without so much as a reading of it. Like most independent thinkers, he was confident that others would come round to his views in the same way he had arrived at them, if only they were permitted to do so.

Closely relevant to this is a point of great interest that is discussed at length in the present book; namely, the cogency of proof that should be possessed by an independent thinker before he ought to urge his views against established tradition. Father Langford himself believes that it was not only imprudent but illogical of Galileo to come forward without demonstrative proofs—proofs which, he points out, had to wait more than a century. But if a man's proofs must be so overwhelming that others will speedily accept them against established authority, then few independent ideas, especially in science, will ever be brought forth, for most really new ideas require the research and the contributions of many men before rigorous proofs are to be found. I think that all that should be required is sufficient weight in the mind of the advocate himself that he will offer himself up to possible general ridicule. Galileo's proofs had at least that much weight for him before he spoke out, and rightly so; his two attempted physical proofs, though not conclusive, were far stronger than many of his critics will allow them to be. I refer to the seasonal variation of sunspot paths and to Galileo's theory of the tides. It would indeed be difficult to explain either of those phenomena without attributing some motion to the earth.

Well, Galileo was not successful with his proofs, and in 1616 the earth's motion was pronounced to be a rash view, philosophically false and contrary to Scripture. That was the opinion of the theological qualifiers called in to decide the point. But the decree then published against certain books did not go so far, as Father Langford has astutely pointed out. The only book to be prohibited outright was not that of Copernicus, but Foscarini's book, which attempted to reconcile the Copernican theory with the Scriptures, and the ban applied only to "others which teach the same." This point, overlooked by nearly all other writers, is essential to an understanding of the later events. Galileo understood it well enough, and his *Dialogue* of 1632 in no way violated the published decree, for he studiously avoided any further attempt (after the unpublished *Letter to Christina*) to show how Copernicus might be reconciled with the Bible. His trial and condemnation hinged on quite another matter than the decree of 1616; namely, his alleged violation of a personal instruction delivered to him in that year.

Now the question whether Galileo did receive an instruction not to hold or defend this view "or teach it in any way, orally or in writing," is the subject of continuing debate because of conflicts among the documents of the case itself. Father Langford believes that Galileo received such an instruction, and consequently that he must have perjured himself in the trial. Many eminent scholars, led in our time by Professor Santillana, are convinced that Galileo never received such an instruction, and view the only concrete evidence that he did as probably a contemporary falsification. It is my own opinion that there was neither falsification by churchmen nor perjury by Galileo; that it was utterly impossible to establish at the trial in 1633 the precise events of 1616, and that those events included an illegal act by an overzealous Commissary of the Inquisition, an act that Galileo had been instructed by a Cardinal to treat as having never happened. The whole affair was, I think, a tragedy of errors. It should now be recognized as one, rather than as the symbol of irresponsible science meddling with society, as Mr. Koestler would have it, or the symbol of an inherent religious intolerance of science, as Professor Santillana maintains.

It is to be hoped that the present candid recognition by a Catholic scholar that the Church's theologians made a mistake in the condemnation of Galileo will open a new era in which writers on this issue, of whatever persuasion, will move toward moderation and objectivity and will show a more rounded understanding of the circumstances which surrounded that great historical drama.

Stillman Drake

INTRODUCTION

The question of Galileo and the Roman Catholic Church seems destined never to die out. Three centuries of myths, prejudiced accounts, and apologetics have distorted the facts and issues of the conflict and made the condemnation of Galileo a subject of enduring prominence. Charges and denials were especially vehement less than one hundred years ago. The spirit of the nineteenth century, with its accent on unbounded intellectual freedom, encouraged historians to produce a graven image of Galileo as the great and courageous scientist whose thoughts were chained by a tyrannical Church. Many historians used Galileo's name as a battle cry in their polemics against the Church of Rome. Catholic scholars, in answering these attacks, often went to the other extreme. Many times what began as an explanation ended as a justification of the condemnation. Facts were overlooked or denied, documents were soft-pedalled, new meanings were concocted for the word "heretical."

The debate continues today. Unfortunately, so does the misunderstanding. An ever-increasing number of books on the history of science are content to repeat the same old charges or their time-worn answers. Critical readers, Catholic and non-Catholic alike, find it annoying to learn that there are still some writers, who, usually in a paragraph or two relating to the famous case, insist on magnifying an admitted mistake into an irretrievable manifestation of unfounded authority and rank stupidity. One can still find those who use the Galileo affair to argue against the doctrinal authority of the pope and to infer that the Church was, and is, a sworn enemy of modern science and human progress.

The old charges are not made as boldly as they once were. In a way, that is unfortunate. It is far easier to answer an argument than it is to dismiss a silent, well-placed implication. For example, even in Professor Wolf's excellent *History of Science, Technology, and Philosophy in the 16th and 17th Centuries*, one reads the misleading statement that:

> The *Dialogue* and other Copernican works remained on the Index until 1822, when at long last the College of Cardinals declared it permissible to teach the Copernican theory in Catholic countries. So the infallible Church had to recant its earlier view.[1]

[1] A. Wolf, *A History of Science, Technology, and Philosophy in the 16th and 17th Centuries* (New York: Harper Torchbooks, 1959), I, p. 37.

Now it is perfectly true that the prohibition against Copernican works was not removed formally from the Index until 1822. But it is not true that only in that year did it become permissible to teach the Copernican theory in Catholic countries. As we will see, Catholics were always allowed to teach it as a theory. Nor was there any question of the infallibility of the Church involved in the Galileo case. This is but one of many examples which could be cited to illustrate the point. It is not unusual to come across references to the condemnation of Galileo in articles and books on various subjects, some as far afield as economics, by authors who presumably have no intention of attacking the Church. But many times their remarks are based on legend rather than fact. And almost always, they miss the real lesson of the Galileo conflict, a lesson which is of paramount importance today.

Perhaps the most influential book on the case today is Giorgio de Santillana's *Crime of Galileo*. It is well-written and based on extensive research. But it often seems to show a lack of historical understanding. As Ernan McMullin, no friend of inquisitorial injustice, has noted:

> The unpleasant polemics between historians of science and Catholic apologists which were so common in an earlier day have more or less subsided . . . but passions still run strong sometimes, even in historians . . . de Santillana's work is one which is likely to perpetuate a shaping of the Galileo symbol one had hoped dead with the nineteenth century.[2]

Intrigue, authoritarianism, ignorance, lies, envy, stubborness: these abounded. But, the *Crime of Galileo* seems to tell us, these, along with falsified documents, were the ultimate and real causes of the conflict. The intellectual issues are swept into a corner. Rarely is there any attempt at a scientific examination of Galileo's proposed proofs for the Copernican system. The scientific, philosophical, and theological problems are too far in the background to be of much value in de Santillana's version of the case. Yet the science, philosophy, and theology of the seventeenth century each had an important part in the condemnation. To slight any one of these areas is to neglect an element necessary for a complete understanding of the famous conflict.

Whenever historical facts are lifted from thier human-temporal context and woven to fit the pattern of preconceived outlooks, one is stretching the facts to fit the thesis. Such a method precludes true and valid understanding of a past event. Human history takes place in unique circumstances. Historical accuracy demands as a prerequisite a thorough knowledge of the people, loves, fears, culture, trends

[2] Ernan McMullen, "Galileo and His Biographers," *The Furrow*, 1960, p. 796.

of thought, and traditions surrounding a past event. Lack of insight into the context of the Galileo case is the glaring defect in most of the writings which deal with it.

Let me say at the outset that this book is not intended to be a Catholic apologia for the decisions of the Holy Office in 1616 and 1633. I am not trying to rob the Galileo case of its relevance. But I do think it is time to correct some of the doctrinal and historical misconceptions which have found unopposed acceptance in much of the literature on Galileo. The condemnation of Galileo was not inevitable. Nor is it the very nature of theology to do battle with modern science. Yet the condemnation is a historical fact. We must examine how it could and did happen and what its meaning is for us today.

My purpose, then, in adding to the already abundant materials published on the condemnation of Galileo by the Church is threefold: first, to establish the facts of the case according to present scholarship; secondly, to get at the meaning of these facts in terms of the scientific, philosophical, and theological issues involved in the conflict, issues which, it will be noted, are still crucial, complex, and disputed today; and, thirdly, to use the facts and issues of the Galileo case to help establish at least a tentative theory of the relationship between science and religion.

If the present work supplies a few insights into the mental milieu of the seventeenth century, if it succeeds in showing the errors that were committed by the principals on both sides in the controversy, errors that must be guarded against today, it will more than have rewarded the years of study and labor that have gone into its preparation. May this book contribute in some way to better understanding among men of good will.

I wish to express my deep gratitude to Father James A. Weisheipl, O.P., D. Phil. (Oxon.), who as professor and friend has been of invaluable assistance in the preparation of this book, and to Mr. Stillman Drake for commenting on the manuscript and contributing the Foreword. I am also indebted to my Dominican brethren, especially the Very Reverend G. J. Graham, Very Reverend B. M. Ashley, and Fathers E. J. Sullivan, T. A. O'Meara, D. P. Morrissey, C. D. Weisser, C. P. Rooney, J. V. Thomas, D. F. Turner, T. J. McCarthy, S. J. Shimek, and B. P. Carey, for their help and encouragement. Generous assistance was also given me by William and Mary Louise Birmingham of Mentor-Omega Books and Mrs. Arabel Porter of New American Library.

J.J.L.

CHAPTER I

The Age of Galileo

The cultural environment and human needs of any age provide both the challenges and the possibilities of success to those chosen few who are destined to alter significantly the course of human history. The greatness, the spirit of any individual who so influences his fellow men can be understood only by those who first see the setting in which he lived and made his contribution. Without some grasp of the cultural, political, social, scientific, philosophical, and theological milieux surrounding a great figure of the past, any attempt to understand the issues which he faced and the content and manner of his response would be doomed to failure.

It is not easy to capture in words the temper of an age. Still the attempt must be made even if the result is imperfect and incomplete. The complexity of factors which foster a typical attitude toward such things as love, learning, life, progress and death does not lend itself to minute analysis or to a neat but sweeping synthesis. Especially is this true of the sixteenth and seventeenth centuries during which Galileo lived. This period has been described with good reason as the "Age of Adventure."

Much like our own day, the Age of Galileo was a time of discovery or rediscovery, of rapid change and anxiety. There were those who considered change as a goal rather than a process, and others who saw the need for true progress and were willing to pay the price to promote it. Then as now, there were long-standing traditions which could not be ignored even though they were bound to be overcome. Some who lived in the Age of Galileo feared excess and desired only that change respect what was essentially unchangeable. But there were also members of that breed of men, still to be found, who seemed to view change as an evil and true freedom as a catastrophe. However one felt about change, he had to admit that it was in the air.

By the dawn of the sixteenth century the average man was probably still somewhat unaccustomed to the startling knowledge that man had conquered the sea and discovered new worlds. The voyage that had been shrugged off for ages as impossible had been navigated by Columbus, and soon armies, such as those of Cortes and Pizarro, would be sent to subjugate peoples whose very existence had always seemed somewhat improbable. Much as children today imagine themselves rocketing to the moon, youngsters of that day probably dreamed of sailing the oceans and discovering new lands which they could claim for their king. As science fiction today speculates on the inhabitants of far-off planets, that age must have had its story tellers who described in great detail the natives of the new world. The impact of Columbus's discovery is greater the more we realize that, unlike the moon or Mars, the New World had actually been touched upon and its inhabitants seen with human eyes.

Portugal, England, and France soon followed Spain in a race to discover and explore. Men willingly sailed unknown seas for God, gold, and glory, though not always in that order. In 1497, John Cabot of England reached Newfoundland. A year later, Vasco de Gama found a sea route around the Cape of Good Hope to India and opened the way for vast trading operations. In 1519 Magellan's ships began a three-year trip around the world.

Jacques Cartier, in 1534, explored the St. Lawrence region for France. These events contributed to a growing spirit of national pride. But it was Spain that led the way. By the middle of the sixteenth century she had colonized large portions of Latin America and royalties were flowing back into the coffers of the Spanish King.

Newly-opened trade markets with the East and the New World hastened the rise of manufacturing techniques, which in turn drew more and more people into the cities and helped weaken the old feudal order. The coinage system began to replace the bartering basket as the accepted vehicle of economy. Bankers in Germany and the Netherlands developed networks of credit which encouraged competitive trade and investment.

Politically, as the Reformation destroyed papal influence in some countries and seriously weakened it in others, the power of the ruling monarch, whether prince or king, was increased. The principle of Machiavelli's *Prince*, that a ruler could use any means he desired in attempting to achieve his political goals, was widely put into practice. The once-powerful Holy Roman Empire was on the verge of collapse. Political alliances were formed and broken almost at will. Thus France during this period was to be allied with and opposed to, at one time or another, the pope, the emperor, the Protestant princes and even the Turks.

Education, which during the Middle Ages had been limited, for all practical purposes, to clerics, nobles, and the wealthy, gradually became more universal. By the sixteenth century it was theoretically possible for students from all classes of society to obtain an education. Italy led the way in the number and importance of its new universities. The Protestant revolt, however, had a harmful effect on the field of learning. The number of students decreased. Church funds set aside for school support were confiscated or diverted to other uses. Many grammar schools attached to monasteries were closed when the monasteries were suppressed by Protestant authorities. In England, for example, there were less than half the number of schools at the end of the sixteenth century as there had been at the beginning. Only after the Council of Trent was education at the pre-university level able to recoup its losses. The Council ordered the establishment of elementary schools and a number of religious communities were established to provide the teachers. The Jesuits, approved by Pope Paul III in 1540, were especially effective in the field of teaching.

The invention and improvement of the printing press was of singular importance in the progress of learning. By the beginning of the Age of Galileo, thousands of works which had been difficult if not impossible to obtain, were made available to scholars and students through the growth of personal and public library collections. New theories which had been refused a hearing in the great universities could now be published and contest established teachings. Each of these factors certainly played a large part in the development of a new age. But it was two movements, the Protestant Reformation and the Renaissance, that made the most critical contributions to the Age of Galileo.

The Reformation was not a sudden outburst against the Catholic Church begun only when Luther posted his 95 theses on the door of the Wittenberg Church in 1517. Nor was it merely a theological break. For decades conditions favoring a revolt had been solidifying. The rise of nationalism, which was opposed to the old theocratic system whereby the papacy was deeply involved in the secular rule of Christian nations, together with a strong desire on the part of rulers to be free of Roman financial and moral interference, explains the welcome which many princes gave to the reformers. In Germany, to take one example, it is estimated that the Church and its clerics controlled approximately one third of the wealth of the entire nation. Some clerics (and even non-clerics) amassed numerous benefices, that is, financial taxes paid to the pastor of an endowed parish. The idea behind benefices was that the pastor would always have a regular means of self support. In return for his labors on behalf of the parish, he was entitled to a sufficient salary. By means of political pull or financial inducements, however, one was able to gain title to several parishes at once and collect the salaries even though he never actually worked in the parish. An old English manuscript tells of a priest and his servant, who, while traveling through England, had the following experience:

> He espied a church standing upon a fair hill, pleasantly beset with graves and plain fields, the goodly green meadows lying beneath by the banks of a crystalline river garnished with willows, poplars, palm trees and alders, most beautiful to behold. This vigilant pastor, taken with the sight of this terrestrial paradise said unto a servant of his: 'Robin,' said he, 'yonder benefice standeth very pleasantly; I would it were mine.' The servant answered, 'why sir,' quoth he, 'it is.' [3]

Church lands were tempting prizes and many princes needed only the incentive and the excuse provided by the Reformation movement to seize them.

The breakdown of papal secular power hastened by the Reformation meant in turn, that the pope could not call upon all the rulers of Europe to crush the revolt in its infancy. Nor could the matter be entrusted to the Holy Roman Emperor. Charles V had enough troubles of his own. For one thing, he and Francis I of France were involved in the early stages of a life-long feud which originated when Charles bribed the electors with money borrowed from the Fuggers of Augsburg and was named Holy Roman Emperor over Francis, who coveted the post. Francis never forgave Charles and he dedicated himself and his armies to the downfall of the Empire. As a result, Charles had to wage a series of wars with France in addition to fighting the advance of the troublesome and powerful Turkish Empire. Small wonder he had little strength left with which to resist the independent land-grabbing of German princes even though he had double reason to do so since he was not only Emperor, but also King of Germany.

But the main issues of the Reformation, as such, were moral and doctrinal. Abuses in the Church were not hard to find. The papacy had several times fallen to men who were unscrupulous

---

[3] Cited by H. O. Taylor, *Thought and Expression in the Sixteenth Century* (New York: Macmillan, 1920), II, p. 60.

or blatently immoral. The clergy was poorly trained and frequently undisciplined. There was a general decline in morals. It is true that there were also good men, holy priests, and dedicated bishops. But their number was all too small.

Added to all this, the doctrinal presentation of Catholic beliefs, after Scotus and Ockham, became detached from its biblical roots, involved in subtle disputations, and dangerously attached to Pelagian presuppositions. These trends undermined the great medieval synthesis of faith and reason masterfully achieved by St. Thomas Aquinas, in which the science of theology was seen as faith seeking an understanding of God's revelation to men, without subtracting in any way from the sublimity or the mystery of the truths revealed. St. Thomas acknowledged that the mysteries of faith could never be understood in this life. But he also showed that man can come, especially through analogy, to a proper, though limited, understanding of the truths which God told mankind about Himself in the Sacred Scriptures. Aquinas was very conscious of the dignity of human nature and one of the basic principles of his theology is that grace does not destroy or debase nature, but rather, perfects it. Nominalism in the fourteenth century attacked the very basis of the Thomistic synthesis. The intramural debates among scholastic schools of theology which followed upon the spread of nominalism led to unnecessarily subtle reasonings and, at times, to useless bickering over philosophy rather than to the study of Bible. The pity of it was that in objecting to the sterile intellectualism which many scholastics appear to have fallen into, some critics sometimes went to the extreme of denying any validity to theological reasoning as a whole. They demanded a non-scholastic return to the printed word of God and a renewed accent on faith as a "leap into the dark." But they also tended to reduce religion to a subjective feeling about God and to deny man's active role in understanding the truths of his faith and his personal cooperation in the attainment of salvation.

Luther, Zwingli, Calvin, and their followers, challenged the

Catholic position on such basic questions as the nature and authority of the Church, the role of the sacraments in the Christian life, the reality of supernatural grace and the freedom of the human will. The necessity of extensive moral reform centered attention especially on the problem of the human will and its part in the working out of salvation. Luther felt that man's nature was depraved and that salvation could come only and solely by faith. Calvin's doctrine of predestination denied man any true causal relationship in the attainment of eternal life. In accord with the accent on the goodness and power of human nature which was revived by the humanists' study of ancient pagan authors, there were those on the other side of the fence who tended to over-accentuate man's role in the spiritual life. It was man's natural strength and goodness which moved him toward God. This view made religion more anthropocentric than theocentric. It was hinted at by Ficino and Pico della Mirandola, among others. Both of these views were in fundamental disagreement with the traditional Catholic teaching that grace is a supernatural gift of God which gives man the power to perform meritorious acts. Efficacious grace moves man's will in accord with his nature, that is, freely, to perform such acts. The philosophical and theological explanations of how the action of grace moves the human will was the subject of a long and heated debate within the Church itself as the Dominicans and Jesuits spent nine years arguing back and forth in the famous *Congregatio de Auxiliis*, to which we will refer later.

There had been criticisms of the ills in the Church prior to the Protestant revolt. The great humanist Erasmus (1466–1536) minced no words in his attacks on the moral and ecclesiastical evils of his times. In the *Praise of Folly* he wrote:

> There are also those who think that there is nothing that they cannot obtain by relying on the magical prayers and charms thought up by some charlatan for the sake of his soul or for profit. Among the things they want are: wealth, honor, pleasure, plenty, perpetual good health, long life, a vigorous old age, and finally, a place next to Christ in heaven. However, they do not

want that place until the last possible second; heavenly pleasures may come only when the pleasures of this life, hung onto with all possible tenacity, must finally depart.

I can see some businessman, soldier, or judge taking one small coin from all his money and thinking that it will be proper expiation for all his perjury, lust, drunkenness, fighting, murder, fraud, lying and treachery. After doing this, he thinks he can start a new round of sinning with a new slate.[4]

Erasmus was no less outspoken with regard to the hierarchy. Of the bishops he wrote:

They believe themselves to be readily acceptable by Christ with a mystical and almost theatrical finery. Thus they proceed with pomp and with such titles as Beatitude, Reverence, and Holiness between blessings and curses—to execute the role of a bishop. Miracles are considered to be antiquated and old-fashioned; to educate the people is irritating; to pray is a waste of time; to interpret Sacred Scripture is a mere formality; to weep is distressing and womanish; to live in poverty is ignominious . . . to die is unpleasant, death on a cross—dishonor.[5]

It has been said that "Erasmus laid the egg which Luther hatched." But while Erasmus and many other thinkers strongly advocated reform, they wanted such reform to take place within the bounds of the Church. Erasmus never forgave Luther for breaking with Rome and inaugurating the scandal of divided Christendom:

I acknowledge Christ, Luther I know not, and I acknowledge the Roman Church, which I hold not to differ from the Catholic Church. From this Church not even death shall tear me, unless she shall openly be torn away from Christ. Sedition I have always abhorred, and would that Luther and all the Germans had felt the same abhorrence.[6]

Luther was to say of Erasmus:

He is a mere Momus, making his mows and mocks at everything and everybody, at God and man, at papist and protestant, but

[4] Erasmus, *The Praise of Folly*, trans. J. P. Dolan, *The Essential Erasmus* (New York: Mentor-Omega, 1964) p. 129.

[5] *Ibid.*, 158

[6] Erasmus cited by J. J. Mangan, *The Life, Character, and Influence of Desiderius Erasmus* (New York: Macmillan, 1927), II, p. 165.

all the while using such shuffling and double meaning terms, that no one can lay hold of him to any effective purpose. Whenever I pray, I pray a curse upon Erasmus.[7]

Catholic attempts at reformation within the ranks of the Church before 1517 had been frequent, but unsuccessful. Influential political and religious leaders were profiting handsomely as things were and they did everything they could to prevent any effective disciplinary measures being taken which would cut into their vested interests. Without the aid and cooperation of Christendom's rulers and bishops, no General Council could be convened, even though several popes had wanted to call such a meeting. The promise of a thorough reform was put off until the Fifth Council of the Lateran (1512–1517), when it was already too late and its decrees were not implemented. To make matters worse, even when the break with the Church was a reality, Pope Leo X seems to have been naively unaware of the tragedy of Luther's revolt and what it signaled for the future. As one Catholic historian has written:

Leo, despite the protests made by the German bishops and several of the universities, issued the bull *Exsurge Domine*, excommunicating Luther, and dispatched Alexander and Eck to Germany to put the bull into execution. Alexander reported to the Pope that Germany was ripe for schism and that disaster was inevitable unless steps should be taken to end the scandals in Rome and to restore discipline throughout the Church. Even then a vigorous campaign of reform might have saved the day; but the Pope was taken up with his art and his amusement, busily engaged in the accumulation of funds to fill his depleted treasury, and more concerned about the political situation than about the spiritual welfare of Christendom. Things drifted from bad to worse.[8]

Once Luther was successful in gaining the allegiance of powerful German princes, the break with Rome was decisive. Church lands were confiscated by armed troops; priests were ordered to adopt the new religion or go into exile. Less than a year after

[7] Luther, *Table Talks*, Hazlitt's translation p. 283, cited by Mangan, *op. cit.*, p. 255.

[8] J. McSorley, *An Outline History of the Church* (St. Louis: Herder, 1943), p. 575.

Luther and his followers gained a solid foothold in Germany, Ulrich Zwingli, in 1519, led the people of Zurich, Switzerland, in a revolt against the Church. Next, John Calvin took up the reform doctrines and by 1535, had won for himself a large following in northern Europe. In 1535, Parliament passed the Act of Supremacy making Henry VIII head of the Church of England, and another nation was lost to the Church. Thus in the short space of twenty years, northern Germany and Scandinavia had rejected Catholicism in favor of Lutheranism, Switzerland and part of France were Calvinist, and England had a national church.

Nearly all of Europe was in a state of turmoil. Hatred, persecution, and wars of religion were to plague the continent for over a century. Protestants and Catholics defended their creeds in bloody conflicts in which no quarter was asked for or given.[9] One wonders if we in our age can really understand the religious confusion, chaos and anxiety which engulfed Europe in the time of Galileo.

Pope Paul III, elected in 1534, overrode the opposition and began in earnest the reform of the Church. He appointed a commission to make a frank and extensive report on ecclesiastical abuses, named new cardinals from men noted for their holiness rather than their wealth, forbade absenteeism on the part of bishops and priests, restricted the granting of indulgences, and established the Congregation of the Holy Office to protect the integrity of Catholic doctrine and morals. But his greatest act was to summon an Ecumenical Council which met at Trent in 1545. Though it was to meet off and on for eighteen years, the Council of Trent provided the impetus and the decrees necessary to bring

[9] Emperor Charles V, seeing that religious unity was no longer possible, approved the treaty of Passau (1552) granting to each prince control of religious matters in his own territory. Three years later, the Peace of Augsburg (1555) confirmed this agreement and the *cujus regio ejus religio* principle was made law. The implications of this principle were drawn to their logical consequence by a number of princes who completely suppressed religious freedom in their domains. This led inevitably to the bitter Thirty Years War which raged from 1618 to 1648 and was settled only with the Peace of Westphalia.

about an effective reform. Trent defined clearly the Church's position on many issues that had been misunderstood or challenged by the Protestants. It also laid down strict disciplinary measures as part of a strong and detailed program of reform.

After the Council, the Church began to regain lost ground. The Dominicans and Jesuits revived Catholic scholarship. A well-organized seminary course was established for training the clergy. Bishops were given more authority over their dioceses. Strict discipline was reestablished in the old religious orders, and new congregations such as the Oratorians, the Theatines and the Brothers Hospitalers were founded. Catholic missions in the New World flourished.

As one might suspect, post-Tridentine theology was, for the most part, apologetic and polemical. Catholic and Protestant theologians rushed into print with defenses, attacks, and counterattacks. The *Centuries* of Magdeburg which attacked the Church with arguments drawn from history was answered by the *Ecclesiastical Annals* of Cesare Baronius. Robert Bellarmine published the influential *De controversiis,* a systematic refutation of current heresies. In addition, Catholics argued with Catholics and Protestants with Protestants on points of doctrine. Giordano Bruno, referring to Protestant theologians, caustically remarks:

> Among ten thousand such pedants there is not one who has not compiled his own catechism, and who, if he has not published it, at least is about to publish that one which approves of no other institution but his own: finding in all others something to condemn, reprove, and doubt; besides, the majority of them are found in disagreement among themselves, rescinding today what they wrote the day before.[10]

As a result of the spirited controversies, theology became a popular subject of discussion in palaces, taverns, ballrooms, and the market place. The common man may not have understood the subtleties very well, but that did not stop him from arguing with gusto for a very definite position. Even more important for

[10] Giordano Bruno, *The Expulsion of the Triumphant Beast,* trans. Arthur Imerti (New Brunswick: Rutgers, 1964), p. 151.

the present study was the influence of the Reformation on the study of Sacred Scripture. Protestants denied the Roman Church's right to be the living interpreter of the Holy Texts. Catholics, in turn, opposed the doctrinal subjectivism that seemed to stem from the principle of private interpretation. Each, in trying to show the validity of its position, focused great attention on the Scriptures themselves. Texts were cited against texts, exegesis against exegesis. Inevitably this led to a conservative, literal attitude toward Holy Scripture: an attitude that was to play a large role in the story of Galileo.

The second general movement that had a major influence on the Age of Galileo was the Renaissance. Generally considered as encompassing the period from 1300 to 1600, the Renaissance did not have a clearly defined beginning or end. Its main feature is that it represented a break with the main currents of the Middle Ages and a preparation for modern secular culture. At one time it was fashionable to think of the Middle Ages as the "Dark Ages" and the Renaissance as a "rebirth" of knowledge. Such a caricature was bound to be erased when more critical methods were employed in historical study and when scholars were able to prove beyond a doubt the depth of intellectual culture to be found in medieval writings and especially those of the thirteenth century. What distinguishes the Renaissance are the objects of its concern and the methods of its scholars. Renaissance interest was primarily centered on the dignity of man and the power of Nature. This interest was excited and sustained by humanism, which, in its strict meaning, was more a method and an ideal than a system of thought. Humanism brought about a revival of ancient learning, both pagan and Christian. As Dr. Paul O. Kristeller, a noted authority on the Renaissance, has pointed out, "humanism was not a new philosophy, but a movement that arose in the field of grammatical and rhetorical studies." It sought to revive the classic writings and it criticized the sterility of style

[11] See Dr. Kristeller's *Renaissance Thought* (New York: Harper Torchbooks, 1961), esp. pp. 100–108.

and inutility of questions raised by nonclassical authors. Humanism opposed as barbarous the scholastic method of the *quaestio*, which stated a point of doctrine in question form, quoted objectives from important thinkers, exposed in clear terms the author's position, and finally answered the objections usually by means of distinctions.

Naturally, the influence of humanism began to be felt in philosophy and theology. Classical studies provided philosophers with accurate texts of a wide range of ancient authors such as Plato, Boethius, Cicero, Aristotle, etc. The revived study of a variety of authors encouraged eclecticism rather than total dedication to any single thinker or system. Study of the pagan writers tended also to renew interest in questions concerned with man, his individual worth, his goals, and his role in the universe.[12] In theology, humanism brought about the serious pursuit of biblical and patristic philology. In other words, the classic textual method was used on the Scriptures and the writings of the Fathers, since these were the Christian classics. It was during the Renaissance that Lorenzo Valla, on the basis of the Greek text, criticized a number of passages in St. Jerome's Vulgate, Erasmus published his famous edition of the Greek New Testament, Cardinal Ximines produced the Complutensian Polyglot Bible, Luther published his popular German translation of the New Testament, and the Douay-Rheims and King James versions were made available.

But the Renaissance was more than an age longing to recreate the glories of the ancient past. It also looked with imagination and originality to the future. Cardinal Nicholas of Cusa (1401–1464), Paracelsus (1493–1541), Cardan (1501–1576), Bernardino Telesio (1509–1588), Giordano Bruno (1548–1600), and others, were liberal thinkers in search of discovery, dissatisfied with what had been done before them. They shared a common

---

[12] Today the term "humanism" in its common usage refers to a concentration on the values, goals, and perfection of man. "Secular humanism" views man as man, independently of God. "Christian humanism" fully acknowledges the dignity of man as man but finds the ultimate cause and goal of human life in God the Creator and Redeemer.

enemy in Aristotelian philosophy even though that system was a powerful force upon the minds of these men who were most outspoken against it. To quote Dr. Kristeller again:

> The authority of Aristotle was challenged during the Renaissance in different ways and for different reasons, but it remained quite strong, especially in the field of natural philosophy. This was due not so much to professional inertia as to the wealth and solidity of subject matter contained in the Aristotelian writings, to which its critics for sometime could not oppose anything comparable. The concepts and methods that were bound to overthrow Aristotelian physics were just being discussed and prepared during the sixteenth century, but they did not bear visible and lasting fruits until the seventeenth ... the Renaissance is still in many respects an Aristotelian age which in part continued the trends of medieval Aristotelianism and in part gave it a new direction under the influence of classical humanism and other different ideas.[13]

In many cases it was not Aristotle himself who was attacked, but Aristotelian thought as it was represented by one or another of the schools that claimed to be his true interpreter. There were, in fact, no less than three fairly distinct brands of Aristotelianism in the sixteenth century. Besides the Thomists who followed Aristotle as interpreted by St. Thomas Aquinas, there was the Averroist school at Padua which adopted the interpretation of the great Arabian commentator and posited one immortal intellect for all of mankind, thereby questioning the personal immortality of the soul and the freedom of the human will.[14] In addition there was a group who can be called the Alexandrists because, on some points at least, they followed the Aristotelianism of Alexander of Aphrodesias. The most eloquent spokesman for this party was Pietro Pomponazzi (1462–1525), who rejected the Thomistic and Averroist notions of immortality and said that it is impossible to prove or disprove the natural immortality of the human soul, though he accepted it as a doctrine of faith.

[13] P. O. Kristeller, *op. cit.*, pp. 46 f.
[14] This was the case at least until the Fifth Lateran Council (1512–1517) condemned those who denied the immortality of the soul, at which time the Christian Averroists revised their position.

The varieties of Aristotelians and their differences were not really very clear-cut. But we must understand the hold that Aristotle had on the universities of Galileo's time and the fact that intramural bickering among the Aristotelians tended to accent textual considerations over experimental evidence and true philosophical vitality. Just as most humanists disliked the scholastic method, there were those who opposed this textual approach to philosophy and who had the ingenuity to search out new vistas of thought. Nicholas of Cusa said boldly that the earth moves and that the universe has no definite bounds. Cardan, Telesio, and Campanella saw the world as an animate organism and they so emphasized immediate sense experience that they attributed the power of sensation to every material thing in creation. Giordano Bruno taught that the universe is a multitude of solar systems existing in limitless space. The earth, he said, was certainly not the center of the universe.

These and other important Renaissance philosophers stressed Nature as a manifestation of the Divine. Its mysteries could be probed by empirical investigation but it was still necessary to speculate on the interconnection of everything in creation. Here there was room for creative and imaginative thinking. The Renaissance respect for the hidden power of Nature encouraged the practice of magic and astrology, and either of these could develop into a serious and profitable profession for anyone who was willing to disobey Church prohibitions and risk punishment by civil authorities.[15]

Thanks to one of the most fascinating characters of the Renaissance, Theophrastus Bombast von Hohenheim, commonly known as Paracelsus (1493–1541), the science of medicine took on new dimensions. He was typical of the age in his awe of nature and an innovator in his insistence on first-hand evidence. Objecting to

[15] For an idea of how widespread the practice of magic was at this time, see Lynn Thorndike, *A History of Magic and Experimental Science* (New York: Columbia U. Press, 1958), XII, and D. P. Walker, *Spiritual and Demonic Magic from Ficino to Campanella* (London: The Wartburg Institute, 1958).

the reverence paid to Galen and Avicenna on medical questions, Paracelsus burned their books publicly in a pan of sulphur and declared that those two authors, long since deceased, were at that very moment suffering a similar fate. While Paracelsus could not resist the gnostic attractions of alchemy and astrology, he made a great contribution by putting across the point that medicine had to consider the whole man and not merely his diseased parts. He insisted that doctors had to use psychological as well as physical treatments to effect a complete cure. This was not the only advance made in medical science during the Renaissance. Leonardo da Vinci (1452–1519) was not only a great artist, he was a gifted scientist as well. He wrote a treatise on the circulation of the blood, although it was not until almost a century later, in 1615, that this circulation was correctly analyzed by William Harvey. His drawings of the human anatomy foreshadowed the important observations and experiments made by Andreas Vesalius in his *De fabrica humani corporis,* published in 1543.

Astronomy too was changed by a man who dared to be an independent thinker. It is to Nicholas Copernicus (1473–1543) that we owe the first thorough formulation of the heliocentric system which came to be known as the Copernican theory, and which Galileo was to adopt with great enthusiasm and defend at tremendous personal cost.

The sculpture and architecture of the Renaissance imitated the classic style of ancient Greece but also introduced the new and often extravagant Baroque. The paintings of Michaelangelo, Raphael and Titian expressed the force, the pride and optimism of the Age.

In nearly every field, the seventeenth century had embarked on an adventure of progress. There was to be no turning back. The philosopher Cardan appreciated this. "Among natural prodigies," he enthused, "the first and rarest is that I was born in this Age." It was into this time of transition, of so many trends and countertrends, of struggle between the old and the new that Galileo Galilei was born on February 15, 1564, at Pisa, Italy. Our knowledge of Galileo's early life is derived mainly from a biography written by his disciple Viviani and published in 1654. We know that the family name had originally been Bonajuti but it had been changed to Galilei several generations prior to Galileo in honor of Galileo Bonajuti, a famous physician.

Galileo's father, Vincenzio, was a hardworking, though not very successful musician. The practical difficulties involved in supporting four children convinced him that his son should follow a more lucrative profession. He wanted the boy to receive a good fundamental education and then to enter the renowned school of medicine at the University of Pisa. At the age of twelve Galileo began his formal education with the monks at the Monastery of Vallombrosa, near Florence. For about two years he concentrated on Latin, Greek, and logic. Impressed by the monastic way of life, Galileo then seems to have joined the Vallombrosan Order as a novice.[16] His father, however, removed him before he had completed his year of novitiate.

In 1581, at the age of seventeen, Galileo began his preparatory study leading to a degree in liberal arts as a prerequisite for the study of medicine at the University of Pisa. But medicine was not to be his vocation either. His native genius in mathematics and mechanics quickly asserted itself and he dedicated himself almost exclusively to these sciences. After four years at Pisa he could no longer afford a formal education and had to leave the University without completing his studies or obtaining a degree. But his intellectual spirit was undaunted by this financial necessity. He continued to study on his own. In 1586, at the age of twenty-two, Galileo invented a hydrostatic balance. Two years later, an able treatise on the center of gravity in solids won him some acclaim.

[16] Arthur Koestler in *The Sleepwalkers* (New York: Macmillan, 1959), p. 354, Angus Armitage in *The World of Copernicus* (New York: Mentor Books, 1961), p. 146, and F. S. Taylor in *Galileo and the Freedom of Thought* (London: Watts, 1938), p. 13, wrongly assert that Galileo was a novice in the Society of Jesus. Research seems to indicate that there was no Jesuit house in Vallombrosa at that time. Galileo did, however, enter the novitiate of the Vallombrosan Order. See F. Horridge, *The Lives of Great Italians* (London: Unwin, 1900), p. 355.

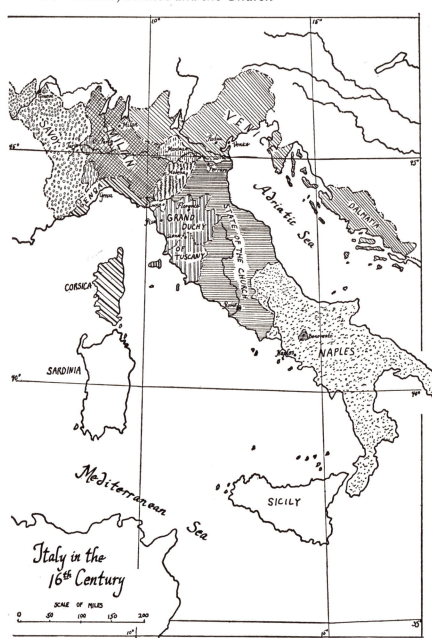

Italy in the
16ᵗʰ Century

Among those who were impressed by his talent was the Marquese Guidubaldo del Monte, himself a mathematician of note.[17] Through the influence of Guidubaldo and his brother, Cardinal Francesco Maria del Monte,[18] Galileo obtained a professorship of mathematics at the University of Pisa in 1589.

The next two years were troubled ones for Galileo. A growing dissatisfaction with Aristotelian cosmology found vivid expression in his lectures. On one occasion, he is supposed to have dropped bodies of different weights from the top of the Leaning Tower of Pisa, showing, since they landed on the ground at the same time, that Aristotle was wrong in believing that the heavier a body is, the faster it will fall. The story, whether or not it is only a legend, exemplifies his spirit.[19] He refused to bow to the authority of any philosopher. He wanted to follow reasoned arguments, not philosophical faith. In the eyes of his fellow faculty members, Galileo was a young upstart, who, though he had not even graduated from the University, dared to challenge them and "The Philosopher."

Much has been written portraying the university Aristotelians of the sixteenth and seventeenth centuries as stubborn, unreasonable philosophers who preferred *a priori* reasoning to direct observation by the senses.[20] But the fact of the matter is that they were not Aristotelians or philosophers in any strict sense. They were products of the lingering Renaissance movement. Their in-

[17] Guidubaldo del Monte was not just an amateur patron of science. He authored a very capable treatise on the science of mechanics entitled *Liber Mechanicorum*, which was first published at Pisa in 1577.

[18] Koestler, *op. cit.*, p. 355, makes Guidubaldo the brother-in-law of Cardinal del Monte. But that they were brothers is clear from a letter written by Guidubaldo to Galileo dated December 30, 1588, in which he says, "*V. S. non lascia occasione di favorirmi, monstrandomi il suo affetto dell' allegrezza che monstra esaltatione del Sr. mio fratello il cardinalato.*" *Le Opere di Galileo Galilei*, Edizione Nazionale, ed. A. Favaro (Florence: 1890–1909), X, 39. Hereafter, this source will be referred to as *Opere*.

[19] It is generally conceded today that Galileo in all probability never performed this experiment at the Leaning Tower of Pisa. See Lane Cooper, *Aristotle, Galileo and the Tower of Pisa* (Ithaca: 1935).

[20] Their position on sense observation has been generally oversimplified. See p. 74, n. 23.

terest in the actual texts of Aristotle, and the absolute authority which they conceded to them, was far from being an Aristotelian or a philosophical approach to reality. In their exaggerated humanism, they were more interested in what the texts of the Stagirite said than in the thoughts or discoveries, however convincing, of their contemporaries. Historians are sometimes too quick in blaming Aristotelian philosophy itself for this lamentable state of affairs. To do so is not altogether fair. Aristotle and the great commentators who followed him through the centuries never pretended to have the final answers to all the problems they considered. It was directly contrary to the spirit of Aristotle to appeal to his writings as the dogmatic criterion of truth. Those who did this were not philosophers. They were philologists. Galileo described them as people who "think that philosophy is a sort of book like the Aeneid or the Odyssey, and that the truth is to be sought not in the universe, not in nature, but by comparing texts!" [21]

The Church had, by the fifteenth century, unofficially adopted Aristotelian philosophy.[22] St. Thomas Aquinas had found many Aristotelian concepts such as matter and form, nature, potency, act, the process of knowledge, and others, to be excellent vehicles for presenting the teachings of the Church. But St. Thomas was careful to distinguish between the realms of faith and reason. Each was in its own right a source of certitude. Each had its limitations. Each was autonomous in its own field of inquiry. Anyone who is familiar with St. Thomas's commentaries on the works of Aristotle knows how careful he was not to dogmatize in philosophy. Dogmatism was not inherent either in the Aristotelian system itself or in the Church's use of this philosophy as a means of

[21] Galileo in a letter to Kepler dated August 19, 1610, *Opere*, X, 423.

[22] By accepting St. Thomas Aquinas's philosophical works, Catholic scholars were, implicitly at least, approving of Aristotle. By the year 1430 the Dominican Hermann Korner could write: "Among all the modern doctors, there is none whose doctrine is so eagerly read, whose books are so closely followed by teachers, and whose writings are so widespread in all the universities of the world as are those of St. Thomas." Cited by J. Berthier, *Sanctus Thomas Aquinas, Doctor Communis Ecclesiae* (Rome: 1914), p. 54.

exposing and defending the truths of faith. Yet between the professors who ruled the universities, and the Churchmen who felt that all of Aristotle's philosophy had to be protected as the quasi-official doctrine of the Church, a stubborn dedication to the status quo set in. The result was a stagnant philosophy whose adherents were so hidebound that they could not tolerate even the suggestion that the system might have minor defects.

Galileo's persistent and often sarcastic attacks on the commonly accepted physics coupled with the unwillingness of his opponents to concede even the smallest point, caused a bitter conflict at Pisa.

Galileo wrote in 1590:

> Few there are who seek to discover whether what Aristotle says is true; it is enough for them that the more texts of Aristotle they have to quote, the more learned they will be thought.[23]

Attacking the Aristotelian explanation of projectile motion, Galileo exclaimed:

> O ridiculous chimaeras! O inept opinions of men, which not only do not approach to the truth, but are opposed to truth itself! But, immortal gods, by what facts, I ask, will it be fitting to believe in these chimaeras of the very men who profess to explain the most hidden secrets of Nature, if in matters, as it were, most open to sense they rashly assert things contrary to the truth?[24]

Such expressions were bound to excite tempers. Added to this, Galileo had incurred the anger of Giovanni de Medici, an illegitimate son of the Grand Duke of Tuscany.[25] As F. S. Taylor narrates the story:

> The illegitimate son of Cosimo de Medici had invented or designed a huge and elaborate dredging machine by which he

---

[23] Galileo, *De motu*, cited by F. S. Taylor, *op. cit.*, p. 39.
[24] Galileo in an unpublished dialogue, cited by F. S. Taylor, *op. cit.*, p. 46.
[25] The Medici family ruled Florence from the year 1429 with the exception of several brief periods when they were temporarily unseated. In 1530, after a brief exile, the Medici with the backing of Spain reconquered Florence. Two years later they transformed the Republic into the Duchy of Tuscany. Within a generation it grew to the status of a Grand Duchy under its Medicean rulers.

intended to clear out the harbor of Leghorn. To design elaborate machinery was a fashion of the period; the idea of magnifying forces by systems of levers, screws, gears, and what-not made the same sort of appeal to the learned amateur as it does today to the schoolboy of mechanical bent. The Grand Duke, to whom the project was submitted, asked Galileo for a report. His opinion is not extant, but we are told that he predicted the failure of the scheme, and we may guess that he did so with his usual emphasis.[26]

The prince seems to have been deeply offended, a consideration which may have helped Galileo decide to resign from the University of Pisa late in the year 1591.

Within a few months, his friend and patron Guidubaldo del Monte recommended him to the Venetian Senate and helped obtain Galileo's appointment to the chair of mathematics at the University of Padua, a post he was to hold for eighteen years. Galileo had to leave his native Tuscany for the Republic of Venice to take the post. But there were compensations. The philosophical atmosphere was more relaxed, and his salary was higher than it had been at Pisa. All during his stay at Padua, Galileo lived in concubinage with a woman named Marina Gamba, though for some reason he never married her. She bore him a son, Vincenzio, of whom not much is known, and two daughters, Polissena and Virginia, both of whom became nuns. It was also at Padua that Galileo invented a thermometer, wrote many treatises on mechanics, studied the physical theories that had developed at the University of Paris, and, most important, became deeply interested in astonomy.

[26] F. S. Taylor, *op. cit.*, p. 47

CHAPTER II

Theories of the Universe

The geocentric theory of the universe, according to which the earth was motionless and at the center of the universe, was the traditionally accepted astronomical system and it enjoyed almost unchallenged favor in the Age of Galileo. Historically, its most influential advocates were Aristotle (378–322 B.C.) and Ptolemy (fl. 150 B.C.). Yet, curiously enough, their systems were quite different. The geocentric theory as taught in the sixteenth and seventeenth centuries was a carefully selected mixture of ideas, incompatible in their proper contexts, but which in combination managed to form a fairly consistent description of the universe. What these ideas were and how they were combined is a necessary and interesting part of our study.

ARISTOTLE

Aristotle's universe was composed of a plurality of real beings that fell into an orderly hierarchy of perfection. Prime matter and substantial form were the principles of every physical body. The simplest bodies occurring in nature were the four elements, earth air, fire and water. These combined to produce the various types of inanimate objects. Living things were more complex bodies, which were united by a higher type of substantial form, called a soul. Aristotle distinguished three types of souls, vegetative, sensitive, and rational, corresponding to the degrees of perfection found in plants, animals, and human beings.

Simple observation of his earthly surroundings showed Aristotle that terrestrial objects were born and died, were changeable and destructible. He further noted that the natural motion of objects on the earth was either up or down. Fire and air being light, naturally moved upward. Earth and water, because they are heavy, move downward toward their natural place in the universe. If a heavy object such as a stone were thrown upward, the motion was "violent"; it was contrary to the natural downward tendency of the stone. As soon as the impetus of the violent motion was exhausted, the stone, striving for its natural place, would descend to the earth.

But in the heavens it was a different case altogether. The celestial world began at the moon. It included the planets, stars, and the transparent, physical spheres in which they were imbedded and which carried them around the earth.[1] Aristotle saw that the motion of celestial bodies was not up or down, but circular. Since no earthly body has a natural capacity for circular motion, there had to be

> some bodily substance other than the formations we know, prior to them all and more divine than they ... it must surely be some simple and primary body which is ordained to move with a natural circular motion as fire is ordained to fly up and earth down ... therefore we infer with confidence that there is something beyond the bodies that are about us on this earth, different and separate from them, and that the superior glory of its nature is proportionate to its distance from this world of ours.[2]

[1] Aristotle does not explicitly say that the spheres are made of solid, transparent matter. But St. Thomas gives five arguments to show that this was really the logic of Aristotle's position. See St. Thomas Aquinas, *In De caelo*, Lib. II, lect. 8, n. 413 (3).

[2] Aristotle, *De caelo*, Bk. 1, Ch. 2, 269a31–269b18.

Thus Aristotle believed that the matter or "stuff" of the celestial bodies differed generically from that of the four terrestrial elements. Celestial matter or the quintessence, as it came to be called, had a number of unique properties. Since circular motion was natural to this special matter and a circle has no limit or contrary, the quintessence must be eternal, immutable, ungenerated, and indestructible. Furthermore, it must be endowed with living, intelligent, and perfect forms. This fits in nicely with Aristotle's ordered hierarchy of perfection. On earth the degrees of perfection ranged from inanimate objects to rational human beings. The higher matter of the heavens was capable of receiving a more perfect form or soul than any on earth. Beyond the outer sphere of the universe there was the most perfect being, the subsisting form without matter, the pure act without any admixture of potency, which for Aristotle was God.

Obviously, Aristotle's cosmology required two quite different systems of dynamics. Everything below the sphere of the moon was governed by terrestrial physics. Everything else in the universe fell under the laws of celestial mechanics. The earth itself had to be immobile. Since the terrestrial elements rise from or fall toward the center of the earth, which is also the center of the universe, the earth must itself be resting in its natural place at the center of the universe. If the movement of the parts of the earth is either up or down, it would be absurd to postulate that the earth itself moves in a circle. It has no need of moving, no place to go. It merely sits quietly as the celestial machine turns wonderously around it. Aristotle remarks that:

> This view is further supported by the contributions of mathematicians to astronomy, since the observations made as the shapes change by which the order of the stars is determined, are fully accounted for on the hypothesis that the earth is at the center.[3]

Aristotle was not an astronomer. But he was interested in formulating a satisfactory explanation for celestial motion. In his work

[3] *Ibid.*, Bk. II, Ch. 14, 279a4–8.

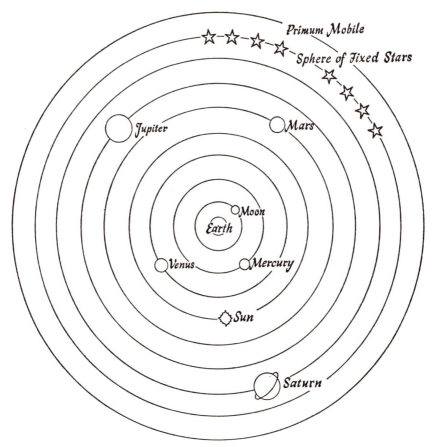

THE ARISTOTELIAN UNIVERSE

De caelo, he accepted the theory of concentric spheres which had been proposed by Eudoxus and elaborated upon by Callippus. According to this system, a squadron of common-centered spheres carried the planets and fixed stars (they were fixed only with regard to their mutual positions) around the central, stationary earth. Astronomers had noted that the planets seem to shift their position from day to day in relation to the fixed stars. At times a planet would stop its direct motion eastward across the sky at a stationary point, and then appear to move westward in retrogressive motion to another stationary point before resuming its eastward direction. We know that the apparent retrogression of Mars, for example, is caused by the fact that the earth travels faster than Mars. Thus when the earth passes Mars, it appears that the planet has stopped and gone in the opposite direction. But Eudoxus found a way to explain the wanderings of the planets while keeping his geocentric system of concentric spheres. He merely assigned several spheres to each planet. Thus in the case of Mars, that planet was attached to the sphere P_1, which was connected at the axis to sphere P_2, which was joined to another sphere, P_3, and P_3 to P_4. By adjusting the tilt and velocity of the four spheres, most of the apparent motions of Mars could be explained. Eudoxus employed twenty-seven spheres and Callippus thirty-three, in accounting for the celestial variations.

Aristotle accepted these calculations and added twenty-one more spheres for a total of fifty-five.[4] The moon, Mercury, Venus, the sun, Mars, Jupiter, and Saturn, each had its own major sphere and its additional nest of helping spheres. Beyond the last planetary sphere was the sphere of the fixed stars, and enclosing the whole universe was the sphere of the *Primum Mobile.* All motion in the universe came ultimately from this primary sphere which was turned by the *Primum Movens Immobile,* the First Unmoved Mover. Each sphere, it is true, had its own unique motion and its own spiritual mover. But the spirits charged with turning the fifty-four inferior spheres were themselves moved by a

[4] Aristotle, *Metaphysics,* Bk. XII, Ch. 8, 1074a12.

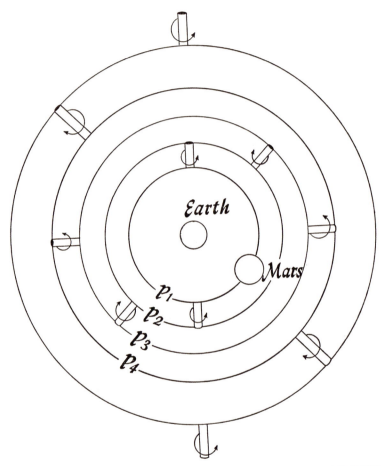

TO ACCOUNT FOR THE VARIOUS MOTIONS OF THE
PLANETS, EUDOXUS AND ARISTOTLE SUPPLIED EACH
PLANET WITH A NEST OF SPHERES. P1 WAS CONNECTED
TO P2, P2 TO P3 AND SO FORTH, EACH SPHERE HAVING
ITS OWN UNIQUE MOTION AND CONTRIBUTING TO
THE IRREGULARITIES IN THE MOTION OF THE PLANET
WHICH THEY WORKED TOGETHER TO MOVE.

desire to conform themselves to the will of the First Unmoved Mover. In the Middle Ages, these notions were Christianized. Angels replaced Aristotle's spiritual movers and the *Primum Movens Immobile,* if taken in the absolute sense, was understood to be God Himself.[5]

Aristotle's system of concentric spheres was soon forgotten. But his doctrines of the quintessence, the circular character of celestial motion, and of the spiritual movers, were to influence mankind for centuries.

PTOLEMY

Claudius Ptolemy (fl. 150 B.C.) agreed with Aristotle in holding that the earth was the center of the universe. But Ptolemy was primarily an astronomer and he was most concerned with solving the mystery of the heavens in the best possible manner. Drawing upon the observations of Hipparchus and the imaginary eccentric circles and epicycles of Apollonius of Perga, Ptolemy, in his greatest work, the *Almagest,* did away with the solid spheres of Aristotle and developed a system which did what astronomy was supposed to do, namely to explain the apparent motions of the heavens, to "save the appearances." [6]

Ptolemy, as did every other astronomer before Kepler, thought the planetary orbits to be perfectly circular, whereas they are in fact elliptical. Thus he too was faced with the task of explaining the puzzling movements of the planets. There were the retrogressive motions which Eudoxus and Aristotle had described by adding spheres. Also, careful observation revealed that the planets seemed to speed up and slow down. And, though earlier astronomers tended to overlook this fact because they could not account for it, the planets varied in brightness. Drawing his ideas from the

[5] See James A. Weisheipl, O.P., "The Celestial Movers in Medieval Physics," *The Thomist,* XXIV, (1962), pp. 286–326.

[6] "Saving the appearances" meant accounting for the apparent motions of the heavens and supplying accurate calculations of the sidereal movements. This could be done with geometric devices which might or might not correspond to the reality which they represented.

work of Apollonius, Ptolemy applied two rather ingenious devices to "save the appearances." First he replaced a number of Aristotle's concentric circles with eccentric circles. In these cases, the planet still circled around the earth, but the earth was no longer at the exact center of the planetary motion. It was off-center. When the planet was nearer to the earth it would appear brighter and seem to change its velocity. A second device which Ptolemy used was the epicycle. On this hypothesis, each planet revolved in a small, tight circle called an epicycle. The center of the epicycle, in turn, revolved around the earth in a large circle called the deferent. This complex of motions could be adapted to explain, with amazing accuracy, nearly all of the celestial phenomena.

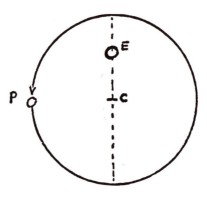

By the use of eccentric circles, Ptolemy could explain the changes in brightness and speed of the planets. As P approaches E, it will appear brighter and faster.

Though the eccentrics and epicycles were regarded in Ptolemy's *Almagest* as mere geometrical devices, they were given a real, physical existence in the work, *Hypotheses of the Planets*, which was authored by someone of the Ptolemaic school, if not by Ptolemy himself. It was a return to the solid spheres of Aristotle. Now each planet was embedded in its epicycle and the epicycle was literally rolled around the earth by the two solid eccentrics which enclosed it. In the twelfth century, when much of Greek thought entered the Latin West through the medium of the Arabs,

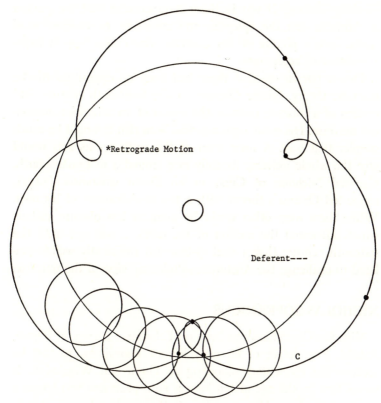

*Retrograde Motion

Deferent---

c

PTOLEMY'S USE OF THE EPICYCLE

IN PTOLEMY'S SYSTEM, THE DEFERENT REVOLVES IN A PERFECT CIRCLE AROUND THE EARTH, WHILE THE EPICYCLE IN TURN IS REVOLVING AROUND THE DEFERENT. LINE C REPRESENTS THE PATH THE PLANET WOULD FOLLOW AROUND THE EARTH.

it was the doctrine of the *Hypotheses,* not that of the *Almagest,* that came with Arab endorsement and, temporarily at least, won favor in the West. A short treatise by John of Holywood (d. 1256) better known as Sacrobosco, recorded a combined Aristotelian-Ptolemaic astronomy. Numerous commentaries on the *Sphere of Sacrobosco* gave the eclectic system popular and scholarly backing. It was in these commentaries that the physics of Aristotle and the epicycles and eccentrics of Ptolemy were woven into a picture of the universe which soon was to enjoy a tradition of acceptance.

This is not to say that there had been no opposition to the geostatic theory. Nicole Oresme, in the fourteenth century, had written of a rotating earth at the center of an otherwise motionless universe. His main argument had been that it would be much simpler that way. A mere turn of the earth on its axis would save the whole universe its daily race around a motionless earth. Cardinal Nicholas of Cusa, in his *Docta ignorantia* (1440), supported Oresme's theory with a few modifications of his own. While there were other works by more or less obscure authors which advocated the motion of the earth, it was not until the sixteenth century that a rival system was sufficiently well-formulated to challenge the Aristotelian-Ptolemaic view of the universe.

NICHOLAS COPERNICUS

Nicholas Copernicus was born on February 19, 1473, at Thorn, Poland. Few details of his early life are known, except that he received his elementary schooling at Thorn and Wloclawek. It seems that by the age of ten he was an orphan and was adopted by his uncle, Lucas Waczenrode, a Catholic priest. In 1489, Uncle Lucas was consecrated Bishop of Ermland, a small principality between Poland, Royal Prussia, and the Gulf of Danzig.

Bishop Lucas took a sincere interest in Nicholas and sent him to the University of Cracow from 1491 to 1495. During these years, Copernicus followed a general course of studies in the humanities. In 1496, after an attempt to have Nicholas appointed a canon of the Frauenberg Cathedral which failed due to some technicalities, the Bishop decided that his nephew should learn Canon Law, since this would equip him for future ecclesiastical posts. The University of Bologna had one of the finest law schools in Europe and it was there that Copernicus enrolled in the fall of 1496. Although his studies at Bologna centered on Church Law, Nicholas also found time for mathematics and astronomy. At Cracow he had met Albert Brudzewski, a noted astronomer who had sparked his interest in the science of the stars. Now, in Bologna, he had the good fortune to know Domenico di Novara, a confirmed Platonist. It was probably Novara who convinced Copernicus that the Ptolemaic system was far too complex to satisfy the principle of mathematical harmony and that there must be a simpler way to "save the appearances." Copernicus was later to record how he searched in the history of thought for a comprehensive and consistent mathematical scheme for charting the heavenly motions:

> Accordingly, when I had meditated upon this lack of certitude in the traditional mathematics concerning the composition of movements of the spheres of the world, I began to be annoyed that the philosophers, who in other respects had made a very careful scrutiny of the details of the world, had discovered no sure scheme for the movements of the machinery of the world, which has been built for us by the Best and Most Orderly Workman of all. Wherefore I took the trouble to reread all the books by philosophers which I could get hold of, to see if any of them even supposed that the movements of the spheres of the world were different from those laid down by those who taught mathematics in the schools. And as a matter of fact, I found first in Cicero that Nicetas taught that the earth moved. And afterwards I found in Plutarch that there were some others of the same opinion: I shall copy out his words here, so that they may be known to all: 'Some think that the earth is at rest: but Philolaus the Pythagorean says that it moves around the fire with an obliquely circular motion, like the sun and the moon. Herakleides of Pontus and Ekphantus the Pythagorean do not give the earth any movement of locomotion, but rather a limited movement of rising and setting around its centre, like a wheel.' [7]

[7] Copernicus, *On the Revolutions of the Heavenly Spheres*, Great Books of the Western World (Chicago: Brittanica, 1952), XVI, p. 508.

In 1597 Bishop Lucas finally succeeded in obtaining his nephew's appointment as a canon of the Frauenberg Cathedral, though he was not formally installed in the Cathedral Chapter until July 27, 1501. Soon after assuming his new post, Copernicus received a leave of absence and returned to Italy, this time to study law and medicine at the University of Padua. Within two years he completed his advanced studies in Canon Law at Padua. He arranged, probably for financial reasons connected with graduation ceremonies and traditions, to receive his Doctor of Canon Law degree from the University of Ferrara. He then returned to Padua and resumed his work in medicine until 1506 when he returned to Ermland as personal physician to his aging and infirm Uncle Lucas.

The six years which he spent in the Bishop's palace at Heilsberg afforded Copernicus the time he needed to study astronomy and experiment with geometrical charts of the heavens. After his Uncle's death in 1512, Copernicus finally moved to the Cathedral at Frauenberg and assumed his duties as a canon. In return for reciting the Divine Office each day together with the other canons of the Cathedral, and for helping administer the temporal properties of the diocese, Copernicus received a regular salary which supported him comfortably. Contrary to a popular belief, Copernicus was not a priest. He probably received the clerical tonsure, but he never advanced to Holy Orders in the Catholic Church.

Copernicus established an observatory at Frauenburg, and, when not occupied as a canon, he was busy at his drawing board. In 1514 Pope Leo X sent out a general request to the heads of many governments and universities urging them to have any experts in theology and astronomy under their jurisdiction submit their opinions on the reforms and corrections necessary in the ecclesiastical calendar. This was to be done either by going to Rome in person or by sending a list of suggestions. At the instigation of Bishop Paul of Fossombrone, Copernicus composed a list of corrections which he sent to Rome.[8]

[8] See Edward Rosen's article "Galileo's Misstatements about Copernicus," *Isis*, XLIX, (1958), p. 319 ff.

Though he had worked out the details of his system much earlier, it was not until 1530 that Copernicus circulated among his friends an outline of his astronomy. The outline or *Commentariolus* immediately attracted widespread attention. In 1533, Pope Clement VII requested Johann Widmanstadt to give a public lecture in the Vatican gardens explaining the Copernican theory. The Pope was quite favorably impressed with it. Nicholas Cardinal Schoenberg wrote to Copernicus urging him to publish, as soon as possible, the complete details of his system. But Copernicus did not want to do so. It was not persecution he feared, but ridicule.[9] He himself admitted, "... the scorn which I had to fear on account of the newness and absurdity of my opinion almost drove me to abandon a work already undertaken."[10]

His fears were not without foundation. As early as 1531 he was satirized in a play staged at Elbing, a few miles from Frauenberg. In 1533, Martin Luther said:

> People give ear to an upstart astrologer who strove to show that the earth revolves, not the heavens or the firmament, the sun and the moon. Whosoever wishes to appear clever must devise some new system which of all systems, of course, is the very best. This fool wishes to reverse the entire science of astronomy; but Sacred Scripture tells us that Josue commanded the sun to stand still, and not the earth.[11]

In 1541 Melanchthon wrote:

> The eyes are witnesses that the heavens revolve in the space of twenty-four hours. But certain men, either from the love of novelty, or to make a display of ingenuity, have concluded that the earth moves ... Now it is a want of honesty and decency to assert such notions publicly, and the example is pernicious.[12]

[9] "The only fear, if it may be so classed, was a shrinking from the ridicule of the unlearned." H. Dingle, *The Scientific Adventure* (London: Pitman, 1952), p. 62. A. D. White in *A History of the Warfare of Science with Theology* (New York: Dover, 1960), I, p. 122, declares that Copernicus feared persecution from the Church. This assertion is considered false by most serious scholars today.

[10] Nicholas Copernicus, *op. cit.*, p. 506.

[11] Martin Luther, *Tischreden*, ed. Walsch, XXII, 2260.

[12] P. Melanchthon, "Initia doctrinae physicae," *Corpus Reformatorum*, ed. Bretschneider, XIII, pp. 216 f.

In 1539 George Rheticus, a mathematician from the University of Wittenberg, came to Frauenberg to study the Copernican system under its author. A year later, Rheticus published a lengthy treatise on the system which was given an enthusiastic reception. Finally, Copernicus agreed to the publication of his manuscripts.

Copernicus dedicated his work to Pope Paul III and then entrusted the text to Tiedemann Giese, Bishop of Culm. Giese sent it to Rheticus, who turned it over to Andreas Osiander, who was to oversee its publication. Osiander, a Lutheran theologian, was well aware that Luther opposed the new system.[13] Wishing to avoid theological difficulties, Osiander wrote an unsigned preface which appeared to be by Copernicus himself and which stated that the heliocentric system was merely a hypothesis intended to serve as a computing device and in no way to represent physical reality. There is little doubt that Copernicus intended his book to represent the real motions of the heavens. As we will see, Osiander's preface was later used to argue that if Copernicus did not consider his system to be representative of physical fact, Galileo should not try to prove that it did.

The work, *De revolutionibus orbium coelestium,* finally reached print. On May 24, 1543, an advance copy was brought to Copernicus who lay on his death-bed. A few hours later the great astronomer died.

In the Copernican theory it is the sun, not the earth, which is

[13] For interesting discussions on the attitudes of leading Protestant reformers toward Copernicus and science in general see E. G. Schweibert, *Luther and His Times* (St. Louis: Concordia Press, 1950), pp. 136 ff. and p. 612, and John Dillenberger's excellent work, *Protestant Thought and Natural Science* (London: Collins, 1961). Professor Dillenberger is of the opinion that "The Reformers did stress the literal or plain meaning of the text, namely what the text said. But what was said could not be equated simply with the information which was provided. The plain meaning of the text was still its theological meaning. The stress on the literal or clear meaning of the text was a weapon against the allegorical interpretation carried on by Rome, an affirmation that the meaning of a text could not be stretched beyond what it actually said. . . . But on one level Luther and Calvin had reservations concerning the place and use of the sciences. Luther was afraid that science might so extend its interest in the natural process of nature to the point where it no longer understood these forces as being under the control of the ever-active sovereign will of God." pp. 31–34.

the center of the universe. The earth is reduced to the rank of a planet having annual, daily, and axial motions. The Pythagorean Philolaus (fl. 520 B.C.), Herakleides of Pontus (fl. 350 B.C.), and Aristarchus of Samos (fl. 280 B.C.) had anticipated, long before, the major elements of the heliocentric system.[14] But Copernicus was the first to construct a true system out of them.

Because he designated all planetary orbits to be perfect circles, Copernicus was forced to adopt eccentrics and epicycles to explain the variations in the celestial motions. By so doing, he undermined the one advantage which his system could have held up against the Ptolemaic, namely, simplicity. As Cohen aptly points out:

> When we consider that Copernicus had to use circle upon circle just as Ptolemy did, then the only major simplification is that the circles needed for the apparent daily rotations of the sun, stars, planets, and moon in the Ptolemaic system could be eliminated on the assumption that the earth rotated daily upon its axis—but almost all the other circles would remain.[15]

Added to this there was the problem that while attempting to stay within the confines of Aristotelian physics, Copernicus's system plainly contradicted a number of Aristotle's basic principles, a step which would never be accepted until a better set of principles could be offered to replace them. Thus the new system was far from the "obvious improvement" some historians have held it to be. Because of the spurious preface, the epicycles, and the lack of an acceptable physics, this system, while it did have a number of supporters through the years, was still considered an inferior hypothesis at the time of Galileo.

Galileo's years at Padua, from 1592 to 1610, were the busiest and happiest of his life. It is probable that during his first two

[14] It was a misnomer to call the heliocentric system "Pythagorean" as was commonly done in the seventeenth century. The Pythagorean school did contribute several general heliocentric concepts, especially in the writings of Philolaus. But Aristarchus of Samos was the true forerunner of Copernicus in antiquity.

[15] I. Bernard Cohen, *The Birth of a New Physics* (Garden City: Doubleday Anchor, 1960), p. 57.

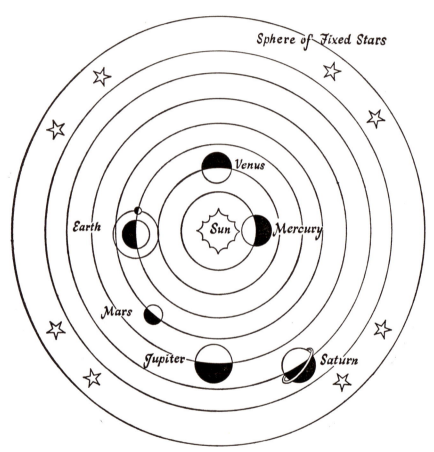

THE COPERNICAN UNIVERSE

years at Padua, he still held and taught the Ptolemaic system. But a growing dissatisfaction with Aristotelian mechanics and cosmology plus a thorough study of the Copernican theory changed his mind. We know from a letter which he wrote to Kepler in 1597 that Galileo had abandoned Ptolemy for Copernicus probably as early as 1594.[16] As A. C. Crombie remarks, it was to become the great passion of Galileo's life to establish the *necessary* truth of the Copernican hypothesis.[17]

A discovery was made in 1608 which revolutionized man's view of the universe. A Dutch spectacle-maker, Hans Lippershey of Middleburg, invented the telescope.[18] A year later, Galileo heard of the invention and began constructing one of his own. In his *Sidereus Nuncius*, Galileo describes how, aided by reports of the Dutch invention, he built a "spyglass" which magnified three "diameters" (*radii*). Next he built a larger, more accurate telescope which he presented to the Republic of Venice, and afterwards, a third one which allowed him to see objects "nearly one thousand times larger and thirty times closer than when seen with the naked eye." [19] With this instrument he searched the heavens.

Important discoveries were not long in coming. He saw that the moon was not a smooth, perfect sphere, as many philosophers expected all the heavenly bodies to be since they were composed of the perfect element, the quintessence. Instead, it was marred with mountains and valleys. Moreover, its earth-like appearance through the telescope seriously questioned the notion that it, and the other celestial bodies, had a material composition generically different from that of the earth.[20] In January, 1610, Galileo dis-

[16] *Opere*, X, 68.
[17] A. C. Crombie, "Galileo's Dialogues concerning the Two Principal Systems of the World," *Dominican Studies*, III (1950), p. 113.
[18] A. Wolf, *op. cit.*, pp. 75 f., gives solid evidence in favor of Lippershey as the inventor of the telescope.
[19] *Opere*, III, part I, 60–61.
[20] This idea of the foreignness of celestial matter was firmly embedded in philosophy of Galileo's time. A contemporary of his wrote that: "The heaven is the first sensible substance because it is composed of matter and form of a different nature than that which makes up inferior bodies. It is first in the genus of physical substance."

covered the satellites of Jupiter. In March of that same year he published a pamphlet entitled the *Sidereus Nuncius* or "Starry Messenger." It contained an account of his discoveries up to the date of publication. The most important discovery was, of course, that Jupiter had its own planets. No longer could it be said without question that all of the celestial bodies revolved around the earth as the physical center of their motions.[21] Here were four planets which had Jupiter as their center of revolution. The satellites also served to answer an objection made against the Copernican theory to the effect that if the earth moved through space, it would leave its moon behind. Here was a planet traveling a major orbit and carrying *four* moons with it.

Galileo dedicated the *Sidereus Nuncius* to Cosimo II de Medici, the Grand Duke of Tuscany, and he named the moons of Jupiter the "Medicean Stars" in his honor.[22] He then asked for and received the post of Ducal Philosopher and Mathematician to Cosimo II. He was given the honorary chair of mathematics at the University of Pisa (undoubtedly a cause of some consternation to his old enemies on the faculty there), but was free from any obligation to teach. He would earn his salary of one thousand gold florins a year by observing the heavens under the aegis of the Grand Duke. He moved to a villa at Arcetri, outside of Florence.

This dictum of Aristotle's is accepted by everyone." Cesare Cremonini, *De calido innato* (Biblioth. Vatic., St. Louis University list 5:33 93). Galileo himself commented that: "... *certo intelligamus Lunae superficiem, non perpolitam, aequabilem, exactissimaeque sphaericitatis existere, ut cohors magna philosophorum de ipsa deque reliquis corporibus caelestibus opinata est* ..." *Opere*, III, part 1, 62–3.

[21] E. J. Dijksterhuis points out that "the discovery that Jupiter had satellites did not in itself constitute a direct argument against the Ptolemaic system—in which, if there was room for motions of planets about points moving in circles [epicyclic motions], motions of satellites about planets could undoubtedly be assumed." *The Mechanization of the World Picture* (Oxford: Clarendon Press, 1961), p. 381. Still it should be emphasized that this was the first time it had definitely been established that one celestial body revolved around a physical body other than the earth.

[22] Cosimo II de Medici succeeded his father as Grand Duke in 1609. Galileo had tutored Cosimo II in Pisa during the summer of 1605.

To the Venetian Republic which had welcomed him eighteen years before, and given him honor, position, and a lifetime professorship, his departure must have looked like desertion. But Galileo was convinced that he needed more free time, uninterrupted by lectures, in order to proclaim the new astronomy and physics which he was in the process of formulating.

Galileo's *Starry Messenger* sold out almost as soon as it was printed. Liberal-minded intellectuals saw it as a great contribution to human knowledge. But the atmosphere was not all peace and triumph. The university Aristotelians, in defense of their Master's cosmology, came forth with angry reclamations. There were those who claimed that Galileo's telescope created illusions. Stillman Drake notes that a leading philosopher, Lodovico delle Colombe, could not believe that the moon was not a perfect sphere. He suggested that the mountains and craters which Galileo had observed were covered over with a smooth, transparent substance, so that the outside of the moon was really as round and smooth as a ball of polished glass.[23] One of the foremost philosophers at Pisa, Giulio Libri, is said to have refused even to look through the telescope.

But there were less pedantic Aristotelians who based their opposition on several strong points. First, they said, the Copernican theory seemed to contradict all direct experience. Anyone with the sense of sight could watch the sun rise in the morning, proceed overhead at noon, and set in the evening. One had only to see the planets at night to be assured that the earth certainly did not look like that! Secondly, they pointed out, the claim that the earth moved was really nothing more than an unfounded assertion which contradicted the accepted system of physics. For example, Ptolemy had opposed the daily rotation of the earth by asking what effect such motion would have on the air surrounding the

[23] Drake, *op. cit.*, p. 73. Dr. Vasco Ronchi, a leading historian of optics, has pointed out that much of the skepticism toward Galileo's discoveries came from the fact that the first telescopes were very imperfect instruments. See Ronchi's *Il cannocchiale di Galileo e la scienza del seicento* (Turin, 1958).

earth. He said that because of the great size of the earth, for it to make a complete turn in twenty-four hours would require such tremendous speed that the light air around it would never be able to keep pace. The air, birds, clouds, etc. would be left behind.[24] Also the opponents of the heliocentric theory appealed to their basic positions on the nature of celestial matter and the distinction between natural and violent motions. In short, they felt that anything which was so diametrically opposed to all the known laws of physics simply could not be true. They had a point. A superior system of physics had to be established before Copernicanism could win acceptance.

Another objection to the Copernican system was that, taken as a fact and not merely as a hypothesis, it would come into direct opposition to certain passages of Scripture that clearly stated the immobility of the earth. This objection was not new. It had been used long before Galileo's time. Luther had quoted the book of Josue against Copernicus. John Kepler's forced departure from the Lutheran faculty at Tubingen was partly due to his Copernican convictions.[25] The learned Jesuit mathematician Christopher Clavius had written in his *Commentary on the Sphere of Sacrobosco* that the new theory seemed to be opposed to the Holy Texts.

Considered as a hypothesis for "saving the appearances" that theory would not contradict Holy Scripture in that it would be understood to be a strictly mathematical device which in no way proposed the motion of the earth as something real but merely as

[24] Ptolemy, *The Almagest*, Great Books of the Western World (Chicago: Brittanica, 1952), XVI, p. 12. The *De magnete* of William Gilbert, published in 1600, had an answer to this problem. But it went largely unnoticed by the men who could have profited most from it.

[25] John Calvin is also frequently accused of anti-Copernican sentiments. However, Dr. Edward Rosen has shown that Calvin never made the remark attributed to him, "Who will venture to place the authority of Copernicus over that of the Holy Spirit?" See Rosen's "Calvin's Attitude toward Copernicus," *Journal of the History of Ideas*, XXI, n. 3 (July, 1960), pp. 431–441. Dr. Rosen defends his thesis against the objections of Dr. J. Ratner in the same journal, XXII, n. 3 (July, 1961), pp. 382–388.

a convenient invention that would aid astronomers in charting stellar positions.

But it was Aristotle himself who was responsible for the strongest argument against the heliocentric hypothesis. If that theory were true, the Philosopher said, stellar displacements or parallaxes would be observable. But none had ever been recorded.[26] The strength of the parallax argument was based on the fact that if the earth moved around the sun, then by first viewing the position of a designated star from the earth's position of the sun, and then six months later from the other side, one should be able to measure a slight shift in the position of the star. This shift would be the star's parallax. Scientifically, the objection was quite valid. Copernicus tried to answer it by saying that the stars are too far away to allow the minute observation necessary to measure a parallax. Kepler saw the strength of the objection and said, in effect, that it was a huge mouthful for the Copernicans to swallow. In point of fact, it remained unanswered until Bessel and others determined the parallax of star 61 Cygni in 1838.

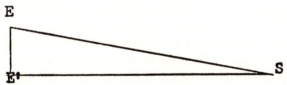

A PARALLAX IS THE APPARENT DISPLACEMENT OF A CELESTIAL BODY BROUGHT ABOUT BY A CHANGE IN POSITION OF THE OBSERVER.

An observer at point E would see a body at S in the direction ES. If he shifted his position to E', the direction would be E'S. The difference between ES and E'S is measured by the angle S and would be the parallax of the body at point S. For bodies near the earth, the length of E E' is taken as the equatorial radius of the earth. For more distant bodies like the stars, since the earth's radius would be too small, the radius of the earth's orbit is used as the distance EE'. Still the stars are so distant that none is known to have a parallax as large as 1″.

[26] Aristotle, *De caelo*, Bk. II, ch. 10, 296b4–7.

Before the debate on his *Starry Messenger* had calmed down, Galileo made two more discoveries: the phases of Venus and the sunspots. The only explanation for the fact that Venus changed its shape just as the moon does was that Venus moves around the sun and not the earth. The sunspots indicated that not all the matter in the heavens was immutable. Yet while Galileo was gathering his proofs and formulating his arguments, he failed to adopt a discovery that might have helped carry the day for Copernicanism. In 1609 John Kepler published his *Astronomia nova* in which he disclosed that the planetary orbits were not circular but elliptical, and that the sun is not at the center but at a focus of the ellipse. This was the very thing needed to clear astronomy of cumbersome epicycles and the dogma of perfect circles. Though it is certain that Galileo knew of the discovery, he never made use of it.

There has been a great deal of speculation regarding Galileo's reasons for assuming a cool attitude toward Kepler. Arthur Koestler is particularly hard on Galileo for his refusal to pay serious attention to Kepler and his unwillingness to answer most of the letters which the German astronomer wrote to him.

But there were reasons why Galileo did not wish to become involved with Kepler. Born on December 27, 1571, at Weil, Germany, John Kepler studied at Tubingen under Michael Maestlin who taught him the Copernican position. Kepler's liberal views in theology and astronomy caused him to leave Tubingen for a post as lecturer in astronomy at Graz. Soon after publishing his first work in 1596, Kepler met Tycho Brahe from whom he received expert training in methods of celestial observation. He succeeded Brahe as Imperial Mathematician, a post he held until his death in 1630.

Throughout his life, Kepler was intrigued by astrology. He was of a somewhat mystical bent anyway and his works are curious mixtures of mathematical brilliance and speculative superstition. It was his connection with astrology, the bizarre nature of some of his ideas, the poverty of his written presentation, as well as his somewhat poetic approach, which kept him from truly influencing Galileo and almost all other astronomers of his own time. Thus while it is easy for us to say that if Kepler's ellipses and Galileo's primitive formulation of the law of inertia had been combined it might have made acceptance of the Copernican system less difficult for the philosophers of the time, it is not hard to understand why these two men were never able to work closely together.

Father Christopher Clavius, chief mathematician and astronomer at the Jesuit Roman College, wrote to Galileo on December 17, 1610, to tell him that the Jesuit astronomers had confirmed his discoveries. Galileo decided that it was time to go to Rome. If he could marshal the support of the Jesuit astronomers and use their endorsement to convince a few influential cardinals that the earth did indeed move, he would have at least unofficial ecclesiastical backing which might help silence the catcalls of his opponents. Full of expectations, he set out for the Holy City on March 22, 1611.

Galileo's reception in Rome was most encouraging. He wrote to his friend Salviati, "I have been received and shown favor by many illustrious cardinals, prelates, and princes of this city." [27] Pope Paul V granted him a long audience. Prince Frederico Cesi named him a member of the recently formed and select *Accademia de' Lincei*, a society devoted to philosophical and scientific studies. The Jesuits of the Roman College were thrilled by his lectures on the stars. They held a day of ceremonies in his honor. When Cardinal Robert Bellarmine asked them for their opinion on the validity of Galileo's discoveries, the Jesuits told him that they were all true.[28] Bellarmine, however, seems to have remained troubled by the scientist's views.

[27] *Opere*, XI, 89.
[28] It was also at this time that Cardinal Bellarmine wrote to the Inquisitor at Padua, "*Videatur an in processu Doct. Caesaris Cremonini sit nominatus ... Galileus, Philosophiae et Mathematicae professor.*" Opere, XIX, 275. This translates as "See whether Galileo, professor of philosophy and mathematics is mentioned in the process of Dr. Cesare Cremonini." Cremonini was in trouble with the Holy Office due to some of his cosmological writings. It is absurd to see anything more than a cautious interest in learning whether

In June, 1611, Galileo returned to Florence satisfied that his name was on the lips of many high Roman officials and that now it was just a matter of time until solid support would come his way.

Despite his stunning findings, Galileo still had no real proof that the Copernican system was anything more than a theory. His observations militated more against Ptolemy and Aristotle than for Copernicus. The appearance of the moon and the sunspots in the telescope did seem to question the distinction between celestial and terrestrial matter. The moons of Jupiter and the phases of Venus did prove that at least some planets revolve around a physical center other than the earth. Elements of the Aristotelian-Ptolemaic system had therefore definitely been shown to be faulty. Had there been only two possible alternatives, either the Ptolemaic or Copernican, Galileo, by discrediting the first would have established the second. But it was not quite that simple. A third system had been worked out by the Danish astronomer Tycho Brahe in 1588. As evidence rendered the Ptolemaic system less and less tenable, the Tychonic conception became ever more prominent.

Tycho Brahe (1546–1601) has rightly been called "the greatest celestial observer of the Renaissance." [30] He rejected the Aristotelian-Ptolemaic universe largely because he was unable to reconcile the supernova of 1572 and the great comet of 1577 with the supposed crystalline spheres and celestial immutability. But he could

---

Galileo was connected at all with the Cremonini business. It hardly measures up to being part of a grand plot, especially since anyone who was aware of the doctrines of Galileo and Cremonini knew that they were diametrically opposed.

[29] Before his death on February 6, 1612, Father Christopher Clavius, the leading Jesuit astronomer, "saw the need of modifying ancient positions by adopting the Copernican position." P. M. D'Elia, *Galileo in China* (Cambridge: Harvard University Press, 1961), p. 85, n. 39. Clavius had confirmed Galileo's discoveries and admitted in writing that astronomers would have to see "how the celestial spheres must be constructed in order to save these phenomena." Clavius, *Opera mathematica*, (Mainz, 1611), III, p. 75.

[30] Grant McColley, "Humanism and Astronomy," *Toward Modern Science*, ed. Robert Palter (New York: Noonday Press, 1961), p. 160.

not accept the Copernican theory either. Despite years of painstaking calculations, he had never been able to discover any parallax. Furthermore, the alleged motions of the heavy and sluggish
earth seemed to be contrary to the principles of sound physics.
Finally, in his opinion, the Copernican system definitely contradicted the words of Sacred Scripture.

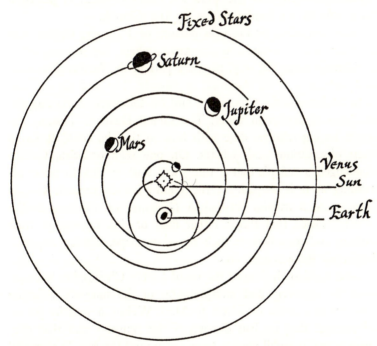

TYCHO BRAHE'S GEOCENTRIC UNIVERSE. ALL THE
PLANETS REVOLVE AROUND THE SUN, WHILE THE
SUN TURNS AROUND THE EARTH.

Since Tycho could accept no previous theory, he constructed
one of his own. In his scheme, the planets revolved around the
sun, while the sun and moon were revolving around a motionless
earth. One can imagine Galileo's frustration, his strong dislike for

the Tychonic system. Galileo saw that it was mechanically pre-
posterous. But he could not prove it as yet. Meanwhile Tycho's
compromise conveniently explained away Galileo's two main discoveries and kept the earth as the motionless center of the universe. The moons of Jupiter and the phases of Venus fit into
Tycho's system just as well as they did into the Copernican. The
Tychonic system was the straw on Galileo's back. It played a
large role in his condemnation by the Church.

After returning to his villa outside of Florence, Galileo was
involved in several minor controversies, which, like so many other
incidentals along the way, played a part in the further alienation
and bitterness of his doctrinal opponents. The first dispute was
with Lodovico delle Colombe [31] and his followers concerning the
behavior of bodies in water. The Grand Duke of Tuscany enjoyed having scholars dine with him so that he could listen to
their observations and debates. On one such occasion, Galileo
challenged a remark that a thing will float or sink in water according as it is shaped. At the Grand Duke's request, Galileo
later wrote a treatise entitled A *Discourse on Floating Bodies*
which showed that things either float or sink because of their
density and regardless of their figure or shape. Galileo won this
battle but the war went on.[32]

Next Galileo challenged the Jesuit Father Christopher Scheiner
over the discovery and interpretation of the sunspots. Scheiner
claimed in a series of letters to Mark Welser, a wealthy patron
of science, that he had discovered the sunspots and that they
were in reality small planets traveling around the sun. Welser had
the letters printed and sent copies to Galileo. Galileo wrote back

[31] Colombe was the leader of the anti-Galileists and his group came to be
called by Galileo's friends the *Columbi* or pigeons. Colombe was a prime
example of the pedantry that opposed Galileo in the name, though certainly
not the spirit, of Aristotle.

[32] Actually, the idea that bodies sink or float according to their shape is
not Aristotelian at all. Aristotle said on this subject that "the shape of bodies
will not account for their moving upward or downward in general, though
it will account for their moving faster or slower." *De caelo*, Bk. IV, ch. 6,
313a15.

to Welser saying that he, not Scheiner, had discovered the sunspots, and that they were not planets, but spots on the sun's surface. As Drake points out:

> ... it is most unfortunate that such a debate should have arisen, as neither man was first to observe sunspots—a phenomenon that was certainly mentioned in the time of Charlemange, and possibly was referred to by Virgil—or even first to publish on the subject. That honor belongs to Johann Fabricius of Wittenberg, whose booklet printed in the summer of 1611 seems to have escaped their attention.[33]

It is highly probable that Galileo did observe the sunspots before Scheiner. It is certain that he did so independent of the Jesuit astronomer. At any rate the dispute was one of lasting bitterness. Galileo had made his first enemy among the Jesuit astronomers.

[33] Drake, *op. cit.*, p. 82.

CHAPTER III

Scriptural Objections

While Galileo was writing his *Letters on the Sunspots* to Welser in 1612, the undertones of the scriptural objection to the Copernican system grew progressively stronger. Luther, Melanchthon, Tycho, and Clavius had all questioned how the new astronomy could be reconciled with the Holy Texts. Now Galileo's opponents recognized that Scripture could be a more effective silencer than their attempts at reasoning had been. Lodovico delle Colombe was the first to use the Bible as a weapon directly against Galileo. His treatise, *Against the Motion of the Earth*,[1] circulated in manuscript, cited several texts from Scripture and implicitly dared the Copernicans to try their hand at exegesis. Galileo was aware of the dangers involved. But he could see the opposition backing off of the issues in physics and retreating to the fortress of the-

---

[1] It is true that Francesco Sizi published his *Dianoia astronomica* in 1610, probably before Colombe's treatise was circulated. But Sizi used Scripture to argue against the reality of the Medicean stars only, while Colombe quoted sacred texts against the motion of the earth. That Colombe was the first to cite Scripture directly against Galileo (though he did not mention Galileo by name) is undisputed. See J. Brodrick, *Robert Bellarmine: Saint and Scholar* (Westminster: Newman, 1961), p. 346.

ology. Dangerous as it might be, he felt he would have to pursue.

In November, 1612, Galileo, while staying at the villa of his friend Salviati, who lived in the country a short distance from Florence, heard that he had been publicly attacked in Florence by a Dominican, Father Niccolò Lorini. Galileo sent a note to Lorini demanding an apology. Lorini, as it turned out, would have been perfectly justified in either ignoring the note or telling Galileo to mind his own affairs and to stop listening to idle gossip. After all, what right had Galileo to demand freedom of expression for himself while denying it to others? Lorini was kind enough to write an explanation to Galileo:

> Most Illustrious Sir,
> Please realize that the suspicion that I, on the morning of All Soul's Day, entered into a discussion on philosophical matters and spoke against anyone, is totally false and without foundation. Not only is it not true, but it is not even likely, since I have not gone outside my field [Church History], nor have I ever dreamed of wanting to get involved in such matters. Neither have I talked about them with Mr. Pandolfini or with anyone else. I don't know what foundation there might be for your suspicion, since this whole business never even occurred to me. It is true that I, not in order to argue, but so that I would not seem to be a deadhead, when the subject was raised, said a couple of words just to show that I was alive. I said, as I still say, that this opinion of Ipernicus—or whatever his name is—appears to be contrary to Holy Scripture. But it makes little difference to me, as I have other things to do ...[3]

Galileo apparently was appeased enough that he could write to a friend a few weeks later, with tongue in cheek:

> Here in Florence there is an incompetent conversationalist who has decided to oppose the mobility of the earth. But this good fellow is so unfamiliar with the founder of that doctrine that he calls him 'Ipernicus.' Behold the source and the subject re-

[2] Giorgio de Santillana adds a touch of drama by telling us that "Preaching on All Souls' Day, 1613, he [Lorini] inveighed against the new theories in unbecoming terms ... so the monks were on the warpath after all." (*Crime of Galileo*, 27.) But as Stillman Drake notes, "Lorini had certainly not preached publicly against Galileo as some writers preposterously state." (*op. cit.*, p. 147, n. 3)

[3] *Opere*, XI, 427.

sponsible for the twisting of poor philosophy.[4]

In March, 1613, Galileo published his *Letters on the Sunspots* under the sponsorship of the Lincean Academy. This marked the first time Galileo had actually endorsed the Copernican system in print. Maffeo Cardinal Barberini, the future Pope Urban VIII, was one of many enthusiastic readers who wrote to congratulate the author. Bypassing Latin, the language of scholars, the *Letters* were published in Italian. This meant that the new discoveries and the Copernican idea were now available to anyone who had curiosity and the ability to read.

The growing controversy now became a popular subject of discussion. It was not long before the interest and speculation of experts and amateurs passed from the complex questions of astronomy and physics and came to rest squarely on the scriptural difficulties involved in the new system. People wanted to know how they were to interpret the text of Josue 10:12–13. "Josue prayed to the Lord, and said in the presence of Israel, 'Stand still, O sun, at Gabaon, O moon, in the valley of Aialon!' And the sun stood still, and the moon stayed, while the nation took vengeance on its foes."[5] Josue would hardly command the sun to stand still if it never moved anyway. Also, they wondered how a moving earth and an unmoving sun could be reconciled with the words of the Psalmist, "The Lord is king, in splendor robed; robed is the Lord and girt about with strength; And he has made

---

[4] *Ibid.*, 461.

[5] A typical exegesis of this text in the time of Galileo can be found in the Scripture scholar Cornelius a Lapide, S.J. (1567–1637). Lapide notes that "a Pythagorean interpretation" was extant which explained that Josue was ordering the sun to be still in the sense of "quiet" so that it could hear his voice and command. He rejects this as ridiculous and explains that the text really means "I command you, O sun, and order you to be immobile and fixed above the cities of Gabaon and Aialon so that you can give us light while we conquer our enemies." And he adds, "When the motion of the sun and the moon ceased the flux and reflux of the sea also stopped, for this is caused by the motion of the moon." This last statement is noteworthy since in 1616 and again in 1632 Galileo tried to convince the world that the tides were not caused by the moon's attraction, but by the motion of the earth. C. a Lapide, *Commentarium in Josue, Job, Judicum* (Antwerp: 1664), p. 52.

the world firm, not to be moved" (Ps. 92:1). Or with the statement that God "fixed the earth upon its foundation, not to be moved forever" (Ps. 103:5). The Book of Ecclesiastes states that "the sun rises and the sun goes down: then it presses on to the place where it rises" (Eccl. 1:5). Psalm 18 adds its testimony: "He has pitched a tent there for the sun, which comes forth like the groom from his bridal chamber and, like a giant, joyfully runs its course. At one end of the heavens it comes forth, and its course is to the other end" (Ps. 18:6–7). These are some of the many texts which could be and were quoted against the new astronomy. Obviously it was more than a matter of one or two obscure texts.

In Galileo's time nearly everyone accepted it as a fact that the sun moved around the earth. Copernicus's *De revolutionibus* was known, but the preface claimed that it was a theory, nothing more. In addition, preachers had for a long time used the notion of the earth being at the center of the universe precisely because it was the abode of man who was the special object of God's concern. Then too, the removal of the earth from the center threatened to destroy the finite, compact universe and confronted man with the fears consequent upon saying that the universe itself was infinite. It is not surprising, therefore, that the Scriptures were called in as a witness.

The scriptural passages in question had a tradition of interpretation which pointed to a stationary earth and a moving sun. The theologians thought, in effect, that here the Bible was teaching a point of science. And there was the mistake. It was one made by an amazing number of theologians of that period. The fact is, and the Galileo matter proved it even to diehards, that the Bible does not claim to give a scientific explanation of the universe. When the sacred writers alluded to physical phenomena, they were guided by sensible appearances, not by a desire to teach physical science.

The new astronomy was the subject of discussion at a dinner given by Grand Duke Cosimo II, late in 1613. A Benedictine monk, Benedetto Castelli, a friend and disciple of Galileo and the chief mathematician at the University of Pisa, defended the views of his master in the informal debate. Castelli left soon after the session ended. He had no sooner gone out of the palace when he was called back by the Grand Duchess Christina, the mother of Cosimo II. As a professional theologian, perhaps he would answer the scriptural objection which, Dr. Cosimo Boscaglia had just assured the Grand Duchess, rendered any motion of the earth impossible. The monk returned to the palace and did his best to answer questions and counter objections brought forth from the Scriptures. Later he wrote Galileo a full account of the affair:

> Wednesday morning I was dining at the court when I was asked about the university by the Grand Duke. I gave him a detailed account of things, with which he showed himself well satisfied. He then asked me if I had a telescope. I said yes, and began describing my observations of the Medicean planets made the night before. Madame Christina wanted to know their position and then the discussion moved to the necessity of their being real objects and not illusions of the telescope. Their Highnesses questioned Professor Boscaglia about this, and he answered that no one could deny their existence. I then related all I knew about your discovery and charting of the orbs of these planets. Don Antonio de Medici, who was at the table, seemed well pleased by what I had said. After more talk, carried on quite solemnly, dinner was finally over and I left. I had hardly gotten out of the palace when Madame Christina's servant caught up with me and said that she wanted me to come back inside. Before I tell you what happened next, let me say that while we were at table Dr. Boscaglia had talked to Madame for a while, and though he conceded all the things you have discovered in the sky, he said that the motion of the earth was incredible and could not be, particularly since Holy Scripture obviously was contrary to such motion. So, to continue my story, I went into the chambers of Her Highness and I found the Grand Duke, Madame Christina, and the Archduchess, Don Antonio, Don Paolo Giordano, and Dr. Boscaglia. After some questions about myself, Madame began to argue against me from Holy Scripture. Then, after making a few protestations of my own inability, I proceeded to play the theologian with such confidence and dignity that you would have enjoyed hearing me. Don Antonio took sides with me and this encouraged me so that instead of being awed by the majesty of Their Highnesses, I conducted myself like a knight. I won

over the Grand Duke and his Archduchess, and Don Paolo even came to my assistance with an appropriate quotation from Scripture. Only Madame Christina kept arguing against me but her manner indicated that she did this only to hear my replies. Professor Boscaglia never said a word.[6]

When Galileo received this letter from Castelli, he decided that the time had come to meet the scriptural challenge head on. He wrote a long letter to Castelli stating his views on science and the Bible. Arthur Koestler writes that while the purpose of this letter was to silence all theological objections to Copernicus, "its effect was the opposite; it became the principal cause of the prohibition of Copernicus and the downfall of Galileo." [7]

Even when copies were made of Galileo's *Letter to Castelli* and were circulating freely, the battle was not yet completely in the open. But the lines were being drawn up. Theologians and laymen were taking sides.

The die was irretrievably cast on December 20, 1614. On that day a Dominican Friar, Father Tommaso Caccini, from the pulpit of Santa Maria Novella in Florence, preached a sermon which strongly condemned the idea of a moving earth. It was one of a series of sermons on the Book of Josue and the topic provided him with his theme. It is often said that he used for his opening text a quote from the Acts of the Apostles (1, 11) "Ye men of Galilee, why stand you looking up to heaven?" Even if this was not his opening quote, it might as well have been. Caccini, a troublesome, ambitious man,[8] left no doubt as to what he thought of the new astronomy or of mathematics in general, for that matter.

Galileo wrote several complaints to friends in Rome letting it be known that he was not at all happy to have been the subject

---

[6] Castelli's letter to Galileo, *Opere*, XI, 605–6. English translation in Drake, *Discoveries and Opinions of Galileo*, pp. 151 f.

[7] Koestler, *The Sleepwalkers*, p. 434.

[8] Caccini is one of the strangest characters in the whole story. G. de Santillana gives convincing evidence that he was tied up with the Pigeon League of Colombe. It seems that the main purpose of his sermon was to force the scriptural issue and silence Galileo. Caccini's behavior stands out in contrast to that of nearly all the other churchmen involved in the case.

of a Sunday sermon. There is no doubt that he had been done a grave injustice. The fact that Father Luigi Maraffi, a Preacher-General of the Dominicans,[9] wrote a formal apology to Galileo did little to placate him and even less to counteract this fresh impetus to theological speculation.

Not long after Caccini's attack, Father Lorini came across a copy of Galileo's *Letter to Castelli*. While three years earlier he had desired to stay out of the controversy, after reading the letter, he changed his mind. He objected to a layman telling theologians how to interpret Holy Scripture. It was one thing for a scientist to speculate about physical matters. It was quite another to write a thesis, however brief, interpreting Scripture to fit those speculations. Father Lorini got a copy of the *Letter* and took it back to his convent. The Dominicans at San Marco's agreed that it looked suspicious. In the back of their minds they saw the possibility that this type of thing could blossom into a full scale practice of private interpretation. Protestant exegesis, even if it did not condone every private interpretation, did at least proclaim as a principle that the Holy Spirit would enlighten every individual who was sincerely searching for the truth which the Scriptures contained for him. The Council of Trent (1545–1563) had decreed in this regard:

> Furthermore, to check unbridled spirits, it [the Holy Council] decrees that no one relying on his own judgment shall, in matters of faith and morals pertaining to the edification of Christian doctrine, distorting the Scriptures in accordance with his own conceptions, presume to interpret them contrary to that sense which holy mother Church, to whom it belongs to judge of their true sense and interpretation, has held and holds, or even contrary to the unanimous teaching of the Fathers, even though such interpretations should never at any time be published.[10]

[9] It is time to correct a mistake made by nearly everyone who has written on the Galileo affair. Luigi Maraffi was never Master-General of the Dominican Order, nor was he *the* Preacher-General. Preacher-General is an honorary title conferred on the outstanding preachers in each Province of the Order. There were probably sixty or more Preachers-General in 1614. Maraffi was not in charge of the Order, but merely wanted to apologize for the excesses of a fellow Dominican. His letter is in the *Opere*, XII, 127 f.

[10] H. J. Schroeder, O.P., *Canons and Decrees of the Council of Trent* (St. Louis: Herder, 1941), p. 18 f.

On February 7, 1615, Father Lorini sent a copy of the *Letter to Castelli* to Paolo Cardinal Sfrondrato, one of the Inquisitors-General in Rome. Lorini explained that:

> All our Fathers of the devout Convent of St. Mark feel that the letter contains many statements which seem presumptuous or suspect, as when it states that the words of Holy Scripture do not mean what they say; that in discussions about natural phenomena the authority of Scripture should rank last; that its exponents have very often erred in their interpretations . . .
>
> When I saw that . . . the followers of Galileo . . . were taking upon themselves to expound the Holy Scripture according to their private lights and in a manner different from the common interpretation of the Fathers of the Church; that they tried to defend an opinion which seemed quite contrary to the sacred text; that they spoke in a slighting way of the Fathers and of St. Thomas Aquinas; that they were trampling under foot all of Aristotle's philosophy, which has been of such service to scholastic theology; and, finally, that, to show their cleverness, they were broadcasting in our Catholic city a thousand impudent and ir-reverent assumptions; when, I say, I saw all of this, I decided to acquaint Your Lordship with the state of affairs, that you, in your pious zeal for the faith, may, together with your illustrious col-leagues provide such remedies as may appear advisable . . . I believe that the Galileists are orderly men and all good Christians, but a little wise and cocky in their opinions . . . [11]

Cardinal Sfrondrato turned the matter over to the Holy Office for examination.[12] An Inquisitor was assigned to read the *Letter* and his report stated that, *"a semitis Catholicae loquutionis non deviat"* [13]—"it does not deviate from the path of Catholic expres-

---

[11] *Opere*, XII, 129.

[12] The Sacred Congregation of the Holy Office was instituted in 1542 by Pope Paul III to defend Catholic faith and morals. While it is generally conceded that there were abuses connected with the Spanish Inquisition, the Roman Inquisition was much milder. The principle upon which it professed to operate was that if criminals could be punished for a crime against human society, heretics, who endangered the eternal salvation of the faithful, were even more deserving of punishment. The procedure was that the accused was given a hearing. If he was judged a heretic, he was given the opportunity to recant and to be restored to union with the Church. If he remained obstinate in his heresy, he was turned over to the civil authorities for punishment. It cannot be denied that although the protection of faith and morals was a desirable goal, the means employed were too often excessive and even theologically unsound.

[13] *Opere*, XIX, 305.

sion." When Galileo heard that his *Letter* had been submitted to the Holy Office, he wasted no time in sending an authentic copy to his friend Archbishop Piero Dini in Rome and asking him to show it to Cardinal Bellarmine. This would insure that he would not be blamed for something he did not write.[14] He sent along a note telling Dini that the *Letter to Castelli* had been written in haste and that he was now revising and expanding it. He finished this task in June, 1615, and entitled the treatise A *Letter to the Grand Duchess Christina*.[15]

While Galileo was composing the revised *Letter*, there was a good deal of activity in Rome. Rumor had it that the Copernican theory was about to be banned.[16] Galileo's friends in Rome went to work pumping the authorities to find out just how things stood. Monsignor Giovanni Ciampoli wrote that he had spoken to Maffeo Cardinal Barberini. He reported, on February 28, 1615:

> Cardinal Barberini, who, as you know from experience, has always admired your ability, told me just last evening that with regard to these opinions he would like to see greater caution in not going beyond the arguments used by Ptolemy and Copernicus, and finally, in not exceeding the limitations of physics and mathematics. The explanation of Scripture is claimed by the theologians as their field, and, if new things are introduced, even by a capable mind, not everyone has the dispassionate faculty of taking them just as they are said.[17]

A few days later Dini wrote to Galileo:

> . . . I had a long talk with Cardinal Bellarmine about the points you mentioned . . . As to Copernicus, his Lordship said that he

14 As it turned out, the copy of Galileo's *Letter to Castelli* which Lorini sent to Cardinal Sfrondrato did contain several variations from the original text. G. de Santillana thus feels that Lorini was involved with those who desired to silence Galileo. I think Koestler is right in saying that the changes were probably not made by Lorini and in pointing out that they made little difference anyway, since the decision of the Holy Office was in favor of Galileo despite the changes in the text.

15 This work, which represented Galileo's theological position was copied and widely circulated, but was not published until 1636 at Strasbourg.

16 This rumor reached Galileo through Castelli who had it straight from the lips of the Archbishop of Pisa.

17 *Opere*, XII, 145–147.

did not believe that his work would be forbidden, and, in his opinion, the worst that could happen to it would be the insertion of a note stating that the theory was introduced in order to save the appearances, or something like that, just as epicycles had been introduced for that purpose. With this reservation, he said, you are at liberty to discuss these matters whenever you wish. Concerning the Copernican system, he said, the greatest obstacle to it seems to be the passage [the sun] "rejoiceth as a giant to run the way" together with the words that follow, which all commentators up to now have understood as implying that the sun is in motion. I replied that Holy Scripture in this passage might simply be using a common mode of expression. He answered that with regard to such a problem we must not be too hasty, just as it would not be right to hurriedly condemn any of these opinions . . . [18]

These two letters together with the Holy Office's quick dismissal of Lorini's denunciation indicate that the acceptability of the new astronomy was still a moot point in Rome.

Within a few days word arrived from Prince Cesi in Rome that a Carmelite friar, Paolo Antonio Foscarini, had just published a book which attempted to show that the Copernican system was not contrary to Holy Scripture. Furthermore, the author had sent a copy to Cardinal Bellarmine and asked his opinion of it. Stillman Drake is perhaps right in saying that this unexpected support from a qualified theologian may have been a crucial factor in Galileo's decision not to accept a compromise.[19] At any rate, Galileo decided to try the ground. He wrote to Dini saying that Copernicus (despite the claims made in the preface of the *De revolutionibus*) had not intended his system as a mere theory and that he, Galileo, wanted it either accepted as a fact or rejected completely. He wrote to Ciampoli to find out how things were going in Rome. On March 19, Ciampoli answered:

[18] *Ibid.*, 151.
[19] Drake, *op. cit.*, p. 161. Foscarini was a very able scholar who had spent six years as Regent of studies at the Carmelite Monastery in Florence and four years as Provincial of the Province of Calabria. He was very interested in the new astronomy and had written a number of treatises dealing with cosmology and mathematics. He died on June 10, 1616, just two months after the sixty-four page book defending the compatability of the new astronomy with Scripture was condemned by the Congregation of the Index.

The great rumors supposed to be circulating here have, I believe, not gone further than to the ears of four or five people at the most. Monsignor Dini and I have tried diligently to discover whether there is anything big in the making, but we have found and heard nothing. Therefore, the report that the whole of Rome was discussing it must be traced to the ones who started the rumors . . . [20]

Cardinal Bellarmine's desire for a temporary compromise was obvious in his reply to Foscarini, dated April 12, 1615. In evaluating the Carmelite's book, Bellarmine wrote:

I have gladly read the letter in Italian and the treatise which Your Reverence sent me, and I thank you for both. And I confess that both are filled with ingenuity and learning, and since you ask for my opinion, I will give it to you very briefly, as you have little time for reading and I for writing.

First. I say that it seems to me that Your Reverence and Galileo did prudently to content yourself with speaking hypothetically, and not absolutely, as I have always believed that Copernicus spoke. For to say that, assuming the earth moves and the sun stands still, all the appearances are saved better than with eccentrics and epicycles, is to speak well; there is no danger in this and it is sufficient for mathematicians. But to want to affirm that the sun really is fixed in the center of the heavens and only revolves around itself [turns upon its axis] without traveling from east to west, and that the earth is situated in the third sphere and revolves with great speed around the sun, is a very dangerous thing, not only by irritating all the philosophers and scholastic theologians, but also by injuring our holy faith and rendering the Holy Scriptures false. For Your Reverence has demonstrated many ways of explaining Holy Scripture, but you have not applied them in particular, and without a doubt you would have found it most difficult if you had attempted to explain all the passages which you yourself have cited.

Second. I say that, as you know, the Council [of Trent] prohibits expounding the Scriptures contrary to the common agreement of the holy Fathers. And if Your Reverence would read not only the Fathers but also the commentaries of modern writers on Genesis, Psalms, Ecclesiastes, and Josue, you would find that all agree in explaining literally (*ad litteram*) that the sun is in the heavens and moves swiftly around the earth, and that the earth is far from the heavens and stands immobile in the center of the universe. Now consider whether in all prudence the Church could encourage giving to Scripture a sense contrary to the holy Fathers and all the Latin and Greek commentators. Nor may it be answered that this is not a matter of faith, for if it is not a matter of faith from the point of view of the subject matter, it is on the part of the ones who have spoken. It would be just as heretical to deny that Abraham had two sons and Jacob twelve, as it would be to deny the virgin birth of Christ, for both are declared by the Holy Ghost through the mouths of the prophets and apostles.

Third. I say that if there were a true demonstration that the sun was in the center of the universe and the earth in the third sphere, and that the sun did not travel around the earth, but the earth circled around the sun, then it would be necessary to proceed with great caution in explaining the passages of Scripture which seemed contrary, and we would rather have to say that we did not understand them than to say that something was false which has been demonstrated. But I do not believe that there is any such demonstration; none has been shown to me. It is not the same thing to show that the appearances are saved by assuming that the sun is at the center and the earth is in the heavens, as it is to demonstrate that the sun is really in the center and the earth in the heavens. I believe that the first demonstration might exist, but I have grave doubts about the second, and in a case of doubt, one may not depart from the Scriptures as explained by the holy Fathers. I add that the words "the sun also riseth and the sun goeth down, and hasteneth to the place where he ariseth, etc." were those of Solomon, who not only spoke by divine inspiration but was a man wise above all others and most learned in human sciences and in the knowledge of all created things, and his wisdom was from God. Thus it is not too likely that he would affirm something which was contrary to a truth either already demonstrated, or likely to be demonstrated. And if you tell me that Solomon spoke only according to the appearances, and that it seems to us that the sun goes around when actually it is the earth which moves, as it seems to one on a ship that the beach moves away from the ship, I shall answer that one who departs from the beach, though it looks to him as though the beach moves away, he knows that he is in error and corrects it, seeing clearly that the ship moves and not the beach. But with regard to the sun and the earth, no wise man is needed to correct the error, since he clearly experiences that the earth stands still and that his eye is not deceived when it judges the sun to move, just as it is not deceived when it judges that the moon and stars move. And that is enough for the present.

I salute Your Reverence and ask God to grant you every happiness.
                    Fraternally,
                              Cardinal Bellarmine [21]
                    12 April 1615

Since Bellarmine's letter to Foscarini was an unofficial but quite definite statement of the Church's attitude toward the new astronomy, we must pause to comment on its major points. The Cardinal urges Foscarini and Galileo to treat the Copernican theory as a hypothesis only. Not aware that the preface to Copernicus's great work was spurious, he thinks that the Canon of Frauenberg never intended it to be taken as a reality. The Jesuits Clavius and Grienberger must have told Bellarmine that the new system saved the appearances better than did Ptolemy's, and he is willing to admit this. But to try to establish the system as fact is asking for trouble. He goes on to quote the Council of Trent and here his theology definitely limps. Trent forbade going contrary to the teaching of the Fathers when their teaching on a subject was unanimous and the subject itself was a matter of faith or morals. Thus for a scriptural interpretation of the Fathers to enjoy unquestionable validity, two requirements had to be met. First, all who wrote on a text had to explain it in the same way. This "unanimous consent" of the Fathers meant that there must be a moral unanimity. If many of the great Fathers interpreted it in one way and no other Church Father contradicted them, the exegesis could be accepted as the universal interpretation of the Fathers. Secondly, the Fathers had to affirm, explicitly or implicitly, that the text under consideration pertained to a matter

---

[21] *Ibid.,* 171 f. Cardinal Robert Bellarmine (1542–1621) was a holy and ascetic man, a brilliant scholar, and one of the greatest theologians of his time. Unfortunately, much of his effort had to be spent writing defenses of Catholic doctrine. Perhaps this is one of the reasons why he did not handle the Galileo affair better. He once remarked that he had spent nearly all his life reading and refuting Protestant writings. Undoubtedly his concern for the faith of the common people gave rise to his very cautious attitude regarding the compatibility of the new astronomy with Sacred Scripture. See J. Brodrick's excellent life of Bellarmine, *Robert Bellarmine: Saint and Scholar* (Westminster: Newman, 1961).

of faith or morals. Therefore, if there was not unanimous consent or if the interpretation was not proposed as a certain doctrine pertaining to faith or morals, but merely as an opinion or conjecture, it did not necessarily have to be followed.[22] Applying this to the case at hand, it would be very difficult to establish that the unanimous verdict of the Fathers was that the sun moved and the earth stood still. Many Fathers of the Church describe the heavens as a Firmament which is unmoved. Nevertheless, on the surface, it would appear that most, if not all, of them did hold that the earth was immobile.

But of the second requirement there can be no doubt. Not one Father can be found who declares that the motion of the heavens or the immobility of the earth pertains to faith or morals. St. Augustine explicitly teaches that it most certainly does not.[23] Next comes Bellarmine's curious distinction between what is a matter of faith *by reason of the subject matter* and what is so *by reason of the one who speaks*. He says that even if the motion of the sun is not a matter of faith from the first, that is, even if astronomy does not fall into the category of faith or morals, it still is a matter of faith because of the second, that is, because it was declared by the Holy Ghost speaking through the Sacred Writers. This is a very poor argument. It begs the question. The issue in question was not the fact of inspiration. Everyone admitted that the Holy Spirit inspired the authors of Scripture with the result that every word was protected from error, every teaching was true. What Foscarini was getting at was that, though we admit that everything which the Scriptures teach is true, we still have to decide what it is that the Scriptures really teach, what it is that they really intend to affirm.

Bellarmine then proceeds to challenge Galileo on scientific grounds. If Galileo can truly demonstrate his theory, then the

[22] See H. Hoepfl, *Tractatus de inspiratione Sacrae Scripturae* (Rome: Biblioteca D'Arte Editrice, 1929), p. 218, and R. C. Fuller, "The Interpretation of Holy Scripture" in *The Catholic Commentary on Holy Scripture* (New York: Nelson, 1953), pp. 60 ff.

[23] St. Augustine, *De genesi ad litteram*, II, 9, 20, (PL XXXIV, 270).

exegetes will have to admit that they were wrong and reinterpret the Scriptures in accord with his demonstration. But this physical demonstration will have to precede any new exegesis. He is asking Galileo for scientific proof. The fact that the new astronomy saves the appearances does not constitute proof that it is the actual system of the heavens. The physicist and historian Pierre Duhem portrayed Bellarmine as a champion of the experimental method.[24] He was not. But he did understand that Galileo's proofs were far from conclusive. Until the new astronomy could show more than a hypothetical advantage over the Ptolemaic, the weight of tradition would have to prevail. In other words, since the heliocentric system had not been demonstrated, Bellarmine felt no need to abandon centuries of philosophical agreement or the interpretation (which, as we know, was really an assumption) of the Fathers. As far as he was concerned, the texts should continue to be interpreted as affirming the motion of the sun and the stability of the earth unless and until human reason might prove that exegesis to be wrong.

The two decisive points of the letter are: first, that the Copernican theory may be treated even as a superior hypothesis, so long as it is proposed as a hypothesis and not as an already demonstrated fact, and, second, that the traditional exegesis of the scriptural passages will remain in force until physical proof can guarantee that this is not the divinely-intended meaning of the texts in question. The exegetical principle is, in itself, quite valid. In fact, it is still approved today:

> In general it may be said that the [Biblical] Commission [of the Catholic Church] holds that the traditional interpretation should be adhered to unless and until, in some particular case and without prejudice to faith and morals, it has been shown that it is more reasonable to maintain some other view.[25]

Obviously the principle in this form includes an important modification. It allows for reasonable probability as sufficient to justify

[24] See Duhem's *Essai sur la notion de théorie physique* de Platon à Galilée (Paris: 1908).
[25] R. C. Fuller, *loc. cit.*

alternative interpretations. Bellarmine demanded a physical demonstration before any change could be made or any other interpretation allowed.

But there was another norm for interpreting Scripture which should have prevented Bellarmine from insisting on too literal an interpretation in this case. The biblical writers did not aim at propounding scientific theory. They had no intention of expounding the hidden mechanisms of sensible phenomena. Their goal was to teach religion, not physics. As the scholarly Cardinal Baronius phrased it, "The Holy Ghost intended to teach us how to go to heaven, not how the heavens go." St. Augustine and St. Thomas Aquinas clearly taught that the Holy Spirit, speaking through the sacred writers, in no way attempted to reveal a system of astronomy. Thus while Sacred Scripture might or might not accurately describe physical phenomena, it was not meant to be used as conclusive proof for or against any purely physical theory. St. Augustine wrote that:

> One does not read in the Gospel that the Lord said: I will send to you the Paraclete who will teach you about the course of the sun and moon. For He willed to make them Christians, not mathematicians.[26]

In commenting on the account of creation as given in the Book of Genesis, Augustine cautioned:

> One could ask which shape and form of heaven must be accepted by faith on the authority of Holy Scripture. Many dispute about these things which the sacred writers passed by in silence, because they are without importance for attaining eternal life ... in short, the Spirit of God which spoke through them did not wish to teach things which contribute nothing to salvation.[27]

St. Thomas Aquinas emphasized repeatedly that in matters of physical science, the sacred writers went by the common conceptions and modes of speech in use among their people. They put down "what God, speaking to men, signified, in the way men

[26] St. Augustine, *De actis cum Felice Manichaeo*, I, 10, (PL XLII, 525).
[27] St. Augustine, *De genesi ad litteram*, II, 9, 20, (PL XXXIV, 270).

could understand and were accustomed to." [28] In questions of this kind, St. Thomas gave two rules to be followed: "First, hold the truth of Scripture without wavering. Second, since Holy Scripture can be explained in a multiplicity of senses, one should adhere to a particular explanation only in such measure as to be ready to abandon it if it be proved with certainty to be false; lest Holy Scripture be exposed to the ridicule of unbelievers and obstacles be placed to their believing.[29] Elsewhere, he remarked that:

> When philosophers are agreed upon a point, and it is not contrary to our faith, it is safer, in my opinion, neither to lay down such a point as a dogma of faith, even though it is so presented by the philosophers, nor to reject it as against faith, lest we thus give to the wise of this world an occasion of despising our faith.[30]

Looking back, we can see that two principles should have been considered by the theological judges of the new astronomy. First, the traditional interpretation was to be held unless solid reasons dictated otherwise. Second, in matters of pure physical science, the Scriptures are not the criterion for establishing one system or forbidding another, since they do not teach science.[31] The correct theological procedure would have been to combine these two principles into a practical and valid norm for solving what appear to be discrepancies between Scripture and science. Had this been done, the opinion that the Scriptures confirmed the sun's motion would have been held as *more probable* even after Galileo's discoveries. By staying within the realm of probability, there would have been room left for another interpretation which would have been permissible, though *less probable*, namely, that the Scripture texts in no way represented scientific affirmations and thus were

---

[28] St. Thomas Aquinas, *Summa theologiae*, I, q. 70, art. 1 and 3, also cited by Pope Leo XIII in his encyclical letter, *Providentissimus Deus*.

[29] St. Thomas Aquinas, *ibid.*, q. 68, art 1.

[30] St. Thomas Aquinas, *Opusculum IX*, "Responsio ad Magistrum Johannem de Vercellis, de articulis XLII," *Omnia Opera St. Thomae Aquinas*, ed. Parma, 1950, vol. XVI, pp. 163 ff.

[31] This is in accord with the procedure recommended in the encyclical of Pope Leo III, *Providentissimus Deus*, which allows Catholic scholars great freedom in proposing interpretations which have reasonable foundations but which also urges them to avoid Concordism.

irrelevant to the scientific question. As proof for the Copernican hypothesis developed, the idea that Scripture was not scientifically authoritative would gradually gain recognition until it became the more probable view and won acceptance as such. In this way, Scripture would not be reinterpreted to meet every new scientific hypothesis that came along, nor would science be forced to bow before an unjustified commitment of theologians to a text which was not meant to teach science.[32]

Bellarmine's mistake was to insist that the only acceptable meaning of the texts was that they affirmed the motion of the sun and the stability of the earth as scientific facts unless a physical demonstration could be produced to force a reconsideration. Besides the force of tradition, an added impetus to Catholic exegetical conservatism resulted as a reaction to the subjectivism inherent in the Protestant principle of private interpretation. Theologians were understandably sensitive whenever they felt that the authority of Scripture or the rights of the Church as its custodian, guardian, and expositor were being questioned. Catholic theology in the seventeenth century had lost a good deal of its elasticity, creativity and *élan*, and the necessity of clarifying and backing up

[32] It might well be asked what is the truth value of the scriptural passages in question? The great Scripture scholar, M.-J. Lagrange holds that they are neither true or false, since truth lies formally in the judgment of the intellect and when one goes by appearances, one does not make an act of judgment with regard to the thing in itself. See Lagrange *Historical Criticism and The Old Testament*, (London: 1905), pp. 112–3.

A. M. Henry resolves the difficulty by saying that, "In such matters, the biblical writers did not aim at propounding scientific theory . . . Whenever they touch upon such subjects, they state nothing more than the most obvious sensible facts which will always remain incontestable, no matter what theoretical explanations may be in vogue. . . . Their aim is above all religious, not scientific. Consequently, without having had to make this disassociation either expressly or even consciously, there was no identification between the material content of the expressions employed, and the real content of such affirmations made in a definite psychological and literary context.

Such unconscious disassociation is not a special consequence of inspiration. Everyday language also uses many terms without dreaming of affirming on every occasion what they abstractly signify. Everybody says "the sun rises, the sun sets," without, for all that, affirming the real movement of this heavenly body. And this was already true when public opinion still unhesitatingly accepted the geocentric theory." Henry, *Introduction to Theology* (Chicago: Fides, 1954), pp. 50–51.

its doctrinal positions led to a somewhat canonical approach in theological matters. Unfortunately there was no one of the stature of Augustine or Thomas Aquinas in the seventeenth century to remind the theologians that the Bible was not intended to teach science. Yet even this faulty theology did not make Galileo's conflict with the Church inevitable. As long as the Copernican theory was treated hypothetically, it would meet no opposition from the Church. If actual proof resulted from scientific research, then the Church would gladly adopt it. Galileo was asked, in effect, not to teach the heliocentric theory as a fact until he could prove it to be one. He himself admitted that:

> The surest and swiftest way to prove that the Copernican position is not opposed to Scripture would be to show with a multitude of proofs that it is true and that the contrary can in no way be maintained. Thus, since no two truths can contradict one another, this and the Bible would be seen to be, of necessity, perfectly harmonious.[33]

Had he been willing to accept the fact that he had no such proof and that all he could do with the evidence at hand was to question several important aspects of the Ptolemaic system (which did not thereby prove Copernicus right or eliminate Tycho's system), the whole matter might have rested there. But Galileo refused to compromise. He would not teach the heliocentric universe as a hypothesis even though he could not prove it to be any more than that. Here Galileo is commonly pictured as having no choice. He had to carry forth the flame of truth without regard to whom or what it burned on the way. It was not that simple, however. Galileo was convinced that he had the truth. But objectively he had no proof with which to win the allegiance even of open-minded men. It is a complete injustice to contend, as some historians do, that no one would listen to his arguments, that he never had a chance. The Jesuit astronomers had confirmed his discoveries; they awaited eagerly for further proof so that they could abandon Tycho's system and come out solidly in

[33] *Opere*, XII, 184.

favor of Copernicus.[34] Many influential churchmen believed that
Galileo might be right, but they had to wait for more proof.
Galileo was asked by his friends to be cautious. To beat the university philosophers at philosophy was one thing: to challenge
theologians in theology was quite another. Bellarmine had given
him an opening, however narrow it might seem to us, "Prove
your theory and we will change our exegesis, otherwise teach it
as a hypothesis which saves the appearances." [35] Even today scientific honesty requires a distinction between hypothesis and fact.
E. A. Burtt points out that:

> It is safe to say that even had there been no religious scruples
> whatever against the Copernican astronomy, sensible men all over
> Europe, especially the most empirically minded, would have
> pronounced it a wild appeal to accept the premature fruits of
> an uncontrolled imagination, in preference to the solid induc-
> tions, built up gradually through the ages, of men's confirmed
> sense experience. In the strong stress on empiricism, so character-
> istic of present-day philosophy, it is well to remind ourselves of
> this fact. Contemporary empiricists, had they lived in the six-
> teenth century, would have been the first to scoff out of court
> the new philosophy of the universe.[36]

Obviously it is not entirely accurate to picture Galileo as an innocent victim of the world's prejudice and ignorance. Part of the
blame for the events which follow must be traced to Galileo
himself. He refused the compromise, then entered the debate
without sufficient proof and on the theologians' home grounds.

Galileo read Bellarmine's letter to Foscarini and jotted down
some pertinent observations on its contents.[37] He presented his

[34] See P. D'Elia, *Galileo in China* (Cambridge: Harvard Univ. Press, 1960)
for a clear indication of the support which Galileo had from the Jesuit
astronomers.

[35] This, of course, calls into question the highly complex problem of the
truth value of working hypotheses. Ptolemy (in the *Hypotheses of the
Planets*) and Copernicus both thought their theories were descriptions of
reality. For a discussion of the principle of "saving the appearances" see
P. Duhem, *The Aim and Structure of Physical Theory*, trans. P. Weiner
(Princeton: 1954), pp. 40 ff.

[36] E. A. Burtt, *The Metaphysical Foundations of Modern Science* (Garden
City: Doubleday Anchor, 1955), p. 38.

[37] See *Opere*, V, 367–370.

position in 1615 in his *Letter to the Grand Duchess Christina,* which was circulated and copied, but not published until 1636. He began the *Letter* with a review of the violent opposition which his discoveries had brought down upon him:

> ...I discovered many things in the heavens which had not been seen before our own age. Both the novelty of these things as well as some consequences which naturally followed from them and which contradicted some physical notions commonly held among academic professors, stirred up no small number of professors against me.[38]

Next he attributes the scriptural objections to the same professors:

> ... They threw various charges and published many writings filled with vain arguments, and they made the serious mistake of sprinkling these with citations taken from places in the Bible which they had failed to understand properly, and which were far from suitable for their purposes.[39]

> ... First they have endeavored to spread the opinion that such propositions in general are contrary to the Bible and are consequently damnable and heretical. They know that it is part of human nature to support causes which tend to oppress one's neighbor, no matter how unjustly, rather than those by which a man might receive some due encouragement. So they have had no trouble in finding men who would preach, with great confidence, the damnability of the new doctrine. And they even preach this from their pulpits thus doing impious injury not only to that doctrine and its adherents, but to all mathematics and mathematicians in general. Next, becoming bolder and hoping (though vainly) that this seed which first took root in their hypocritical minds would branch out and upward, they began spreading rumors among the people that before long this doctrine would be condemned by the supreme authority.[40]

After reminding his readers that Copernicus's book was accepted by the Church long before this dispute arose, he asks why it should be condemned now, especially after "manifest experiences and necessary proofs have shown the Copernican doctrine

[38] *Letter to the Grand Duchess Christina, Opere,* V, 309 ff. translation in Stillman Drake's *Discoveries and Opinions of Galileo,* pp. 175 ff. References will be to Drake's translation which I am using with a few minor modifications.
[39] *Ibid.*
[40] *Ibid.*, p. 179.

to be well grounded." [41] And, he adds pointedly, Copernicus's book should certainly not be condemned by those who have never even read it:

> I hope to show that I proceed with much greater piety than they do when I argue not against condemning this book, but against condemning it in the way they suggest—that is, without understanding it, weighing it, or so much as reading it. For Copernicus never discusses matters of religion or faith, nor does he use arguments that depend in any way upon the authority of sacred writings, which he might have interpreted erroneously. He stands always upon physical conclusions pertaining to the celestial motions, and deals with them by astronomical and geometrical demonstrations, founded primarily upon sense experiences and very exact observations. He did not ignore the Bible, but he knew very well that if his doctrine were proved, then it could not contradict the Scriptures when they were rightly understood. [42]

Galileo next asks whether it is heretical to hold that the sun is motionless and that the earth moves. First, he points out that the Bible can never err, whenever its true meaning is understood. Since, however, many passages are abstruse and affirm things quite different from what their bare words signify, an exegete who confines himself to the unadorned grammatical meaning might fall into error:

> Not only contradictions and propositions far from true might thus be made to appear in the Bible, but even grave heresies and follies. Thus it would be necessary to assign to God feet, hands, and eyes, as well as corporeal and human affections such as anger, repentance, hatred, and sometimes even the forgetting of things past and ignorance of those to come. These propositions uttered by the Holy Ghost were set down in that manner by the sacred scribes in order to accommodate them to the capacities of the common people, who are rude and unlearned. [43]

Secondly, he points out that whenever the Bible speaks of physical conclusions, the sacred writers were careful to avoid confusing the minds of their readers and rendering them suspicious of the

[41] *Ibid.*
[42] *Ibid.*
[43] *Ibid.*, p. 181.

religious truths which the Bible was meant to transmit. In other words, the Bible does not pretend to teach science:

> Now the Bible, merely to condescend to popular capacity, has not hesitated to obscure some very important pronouncements, attributing to God Himself some qualities extremely remote from (and even contrary to) His essence. Who, then, would positively declare that this principle has been set aside, and the Bible has confined itself rigorously to the bare and restricted sense of its words, when speaking but casually of the earth, of water, of the sun, or of any other created thing? Especially in view of the fact that these things in no way concern the primary purpose of the sacred writings, which is the service of God and the salvation of souls—matters infinitely beyond the comprehension of the common people.[44]

Thirdly, Galileo rightly states that in physical matters one should begin looking for the truth not from the authority of scriptural passages, but from "sense experience and necessary demonstrations." Thus:

> ... nothing physical which sense experience sets before our eyes, or which necessary demonstrations prove to us, ought to be called into question (much less condemned) on the testimony of biblical passages which may have some different meaning beneath their words.[45]

So far so good. He has spoken like a superior theologian. But now he falters:

> Yet even in those propositions which are not matters of faith, the authority of Scripture ought to be preferred over that of all human writings which are supported only by bare assertions or probable arguments, and not set forth in a demonstrative way. This I hold to be necessary and proper to the same extent that Divine Wisdom surpasses all human judgment and conjecture.[46]

And again:

> I take this to be an orthodox and indisputable doctrine, and I find it explicitly in St. Augustine when he speaks of the shape of heaven and what we may believe concerning it. Astronomers

[44] *Ibid.*, p. 182.
[45] *Ibid.*, pp. 182 f.
[46] *Ibid.*, p. 183.

seem to declare what is contrary to Scripture, for they held the heavens to be spherical while the Scripture calls it "stretched out like a curtain" [Ps 103:2]. St. Augustine is of the opinion that we are not to be concerned lest the Bible contradict astronomers; we are to believe its authority if what they say is founded only on the conjectures of frail humanity. But if what they say is proved by *unquestionable arguments,* this holy Father does not say that the astronomers are to be ordered to dissolve their proofs and declare their own conclusions to be false. Rather, he says, it must be demonstrated that what is meant in the Bible by "curtain" is not contrary to their proofs.[47]

He goes on to say:

From the above words I conceive that I may deduce this doctrine: that in the books of the sages of this world there are contained some physical truths which are soundly demonstrated, and others that are merely stated. As to the former, it is the job of wise theologians to show that they do not contradict the Holy Scriptures. As to the propositions which are stated but not rigorously demonstrated, anything contrary to the Bible involved in them must be held to be undoubtedly false and should be proved so by every possible means.[48]

Thus after his praiseworthy defense of the fact that Scripture does not teach science, now he yields. Scripture, he says, is a superior authority even in science unless a contrary physical argument is *demonstrative.* Galileo, in this series of passages at least, makes the same mistake as Bellarmine did in conceding the highest authority to Scripture even in matters which are not of faith or morals. He gives Scripture strict scientific authority over physical arguments which are only probable. And scientifically he loses his case. He had no demonstrative proof of the Copernican system. At the most, all he could show was that the Copernican hypothesis was superior to the Ptolemaic and it was not easy even to do this. Moreover, Tycho's system accounted for all of Galileo's discoveries with as much ease as the Copernican theory did. And Tycho still had the earth as the immobile center of the universe. The issue at hand was not whether Copernicanism was permis-

[47] *Ibid.,* pp. 197 f. Emphasis mine.
[48] *Ibid.,* pp. 194 f.

sible as a hypothesis, but whether it could be upheld as the real system of the heavens. By granting Scripture precedence over probable physical arguments to the contrary, and by his inability to prove the real truth of the Copernican system, Galileo was caught in a logical snare. It was a snare which he could have avoided had he only reaffirmed in the places cited above what he said in the rest of his *Letter*, namely, that Scripture, because it was not meant to teach astronomy, was not admissible as evidence in astronomy.

The *Letter* then proceeds to misrepresent the Church's position on Copernican astronomy. After stating that theologians should not arrogate to themselves the authority to decide on matters which are outside of their competence, Galileo writes:

> Again to command that the very professors of astronomy see to the refutation of their own observations and proofs as if they could ony be fallacies and sophisms, is to command something which is impossible to accomplish. For this would amount to commanding that they not see what they see, and must not understand what they know, and that in searching they must find the opposite of what they actually discover.[49]

Was this really what the Church had asked Galileo to do? Or did Cardinal Bellarmine merely let it be known that Copernicanism was better treated as a hypothesis until more solid proof could be brought forward? Hadn't the Jesuit astronomers confirmed all of his discoveries and studied their implications?

Throughout the *Letter* we can see indications that Galileo was afraid that the Copernican theory might be condemned outright. He had this attitude even though the best sources assured him that, at worst, a few minor changes were contemplated for the *De revolutionibus* which would insure its hypothetical character. We have already cited Dini's letter to Galileo in early March, 1615, which said, "As to Copernicus, his Lordship said that he did not believe that his work would be forbidden, and, in his opinion, the worst that could happen to it would be the insertion of a note stating that the theory was introduced in order to save

[49] *Ibid.*, p. 193.

the appearances...." His friend, Castelli, referring to the rumored condemnation wrote, "Subsequently I have heard to my great satisfaction that the gossip at Rome is not such a great matter as it was said to be. And it appears to me that the rumor made at Rome is not Roman, but alien. I mean that it has been fabricated by these same gentlemen who have likewise produced it at Florence." [50]

When the *Letter to the Grand Duchess* was copied and circulated the dispute entered its critical stage. Many philosophers and theologians must have seethed at what appeared to be an attempt to assert that the heliocentric system was a demonstrated truth and therefore traditional exegesis should bow before it. That Galileo knew it was not demonstrated is obvious in the following passages:

> ... but when nearly everything the philosophers and astronomers say on the other side is proved to be quite false, and all of it inconsequential, then this side should not be deprecated or called paradoxical simply because it cannot be completely proved.[51]
>
> ... As to rendering the Bible false, that is not and never will be the intention of Catholic astronomers such as I am; rather, our opinion is that the Scriptures accord perfectly with demonstrated physical truth. But let those theologians who are not astronomers guard against rendering the Scriptures false by trying to interpret it against propositions which may be true and might be proved so.[52]
>
> Hence I should think it would be the part of prudence not to permit anyone to usurp scriptural texts and force them in some way to maintain any physical conclusion to be true, when at some future time the senses and demonstrative or necessary reasons may show the contrary. Who indeed will set bounds to human ingenuity? Who will assert that everything in the universe capable of being perceived is already discovered and known? [53]

But, reading the *Letter to the Grand Duchess*, one could easily get the impression that its author felt he had already demonstrated

---

[50] Castelli, cited by Drake, *op. cit.*, pp. 157 f.
[51] Galileo's notations on Bellarmine's letter to Foscarini, trans. in Drake, p. 169.
[52] *Ibid.*, p. 168.
[53] *Letter to the Grand Duchess*, p. 187.

the new astronomy. It was not Galileo's intention to deceive his readers, but some of his remarks are openly exaggerated or at least ambiguous. For example:

> They also know that I support this [the Copernican] position not only by refuting the arguments of Ptolemy and Aristotle, but by producing many counter arguments, in particular, some which pertain to physical effects whose causes can perhaps be accounted for in no other way.[54]

Again and again in the *Letter* he talks about the "manifest sense experience and necessary demonstrations" which cannot be condemned on, and need not be subject to, the authority of Scripture. He seems to imply that the Copernican theory is backed by such experience and demonstrations. Perhaps he thought there was no point in giving his arguments in a logical sequence backed up by his experimental observations because he felt that no one would listen even if he did. After all, he had written to Dini:

> How can I do this [give proofs which demonstrate the Copernican position] and not merely be wasting my time, when those Peripatetics who must be convinced show themselves incapable of following even the simplest and easiest of arguments, while on the other hand they are seen to set great store in worthless propositions.[55]

Now that is not an entirely fair accusation. If he had had the proofs, he would have found a great deal of support, especially from the Jesuits. Rather than give a thorough treatment of his proofs in his *Letter to the Grand Duchess*, he claimed that:

> If truly demonstrated physical conclusions need not be subordinated to biblical passages, but the latter must rather be shown not to interfere with the former, then before a physical proposition is condemned, it must be shown to be not rigorously demonstrated—and this is to be done not by those who hold the proposition to be true, but by those who judge it to be false. This seems to be very reasonable and natural, for those who believe an argument to be false may much more easily find the fallacies in it than can those who consider it to be true and conclusive.[56]

[54] *Ibid.*, p. 177.
[55] *Opere*, XII, 183–4, in Drake, p. 166.
[56] *Opere* V, 327, in Drake, pp. 194 f.

Then, to those who employed the Scriptures against him, Galileo issued this challenge:

> . . . knowing that a proposition cannot be both true and heretical, let them employ themselves in the business which is proper to them; namely, demonstrating its falsity . . . if it is impossible for a conclusion to be declared heretical while we remain in doubt as to its truth, then these men are wasting their time clamoring for the condemnation of the motion of the earth and the stability of the sun, which they have not yet demonstrated to be impossible or false.[57]

The issue has become clouded. It must have seemed to the theologians that Galileo was demanding that the Church must accept Copernicanism and change the official interpretation of the scriptural passages, or she must show it to be false and condemn it. Galileo seemed to be leaving no room for considering the Copernican theory even as a superior hypothesis until further proof could be adduced.

Let us review the key points of Galileo's position. He was correct in saying that Scripture does not intend to teach physics or astronomy. But he was wrong in holding that Scripture has absolute scientific authority over probable physical arguments which seem to oppose it. Bellarmine had said, in effect, "If you want us to change the scriptural interpretations, you will have to demonstrate the truth of your theory." He could now say to Galileo, "You admit that scriptural authority exceeds probable physical arguments. So do we. Yet you insist on asserting without sufficient proof that your system is true and our exegesis is wrong. You are being illogical and asking us to follow suit."

Finally, Galileo's demand that before the Church can condemn the Copernican system theologians must prove it to be "not rigorously demonstrated" and prove it to be "impossible or false" looks like a switching of the burden of proof. Now Galileo had no intention of throwing the burden of proof over to the side of the theologians. What he wanted was for the theologians to examine all the arguments in favor of the Copernican position before

[57] *Ibid.*, pp. 210 f.

making any judicial decision. Unfortunately, this is not the impression which he seems to have given. He did not make it clear enough, first of all, that he knew his proofs were not demonstrative, secondly, that the necessary demonstrations of which he spoke would come in the future, and, thirdly, that since science is not learned in the Bible, new exegesis should be possible.

Simply put, the paradox was this: Galileo was right theologically in holding that the Bible is not a science text book. Bellarmine was right "scientifically" in demanding that a hypothesis remain only a hypothesis until proven as a fact. But both Galileo and Bellarmine were wrong in conceding absolute authority to Scripture if contrary physical arguments were only probable. They were both bowing to a black or white alternative. Bellarmine refused to settle for the traditional interpretation as the more probable exegesis pending further proof. Galileo appears to have been unwilling to consider his theory as an unproven hypothesis until he could strengthen his scientific arguments. Therein lie the seeds of the whole unfortunate conflict.

CHAPTER IV

The Decree Against Copernicus

With the issue still up in the air and any important developments bound to occur in Rome rather than Florence, Galileo decided, late in 1615, that he would go in person to the Holy City to see what he could do to quiet some rumors which were supposed to be circulating about him there, and to win the support of Church authorities. The Grand Duke's Ambassador in Rome, Piero Guicciardini, advised against his coming. So did many of his friends in Rome. They thought that the issue was too sensitive right then and that Galileo's presence would not help matters any. But he overruled their objections. It was to prove a serious mistake. Galileo had good assurance that the unofficial Church position was "wait and see" and that no amount of agitation from known characters such as Caccini could alter that position. It was up to Galileo, whether or not he realized it, to make the Church move one way or the other. He could bolster his arguments and gradually compel acceptance of the new astronomy. Or he could force a decision despite his lack of evidence and hope for the best.

Galileo arrived in Rome on December 7, 1615, and again he

received a cordial reception. Perhaps the enthusiasm expressed by some of his followers in Rome flattered him into overestimating his position. He must have found out that the Holy Office had dismissed Caccini's complaint against him. This was his chance and he was not going to pass it up. Now, right in the theologians' front yard, he preached Copernicanism. He went from one house to another explaining, debating, and winning. As fast as Aristotelians could be found to oppose him, he bowled them over with a display of his talent for polemics. To many observers it must have seemed that he was trying to destroy the doctrines of Aristotle and Ptolemy in order to claim a victory for Copernicanism by default. One observer wrote:

> We have here Mr. Galileo, who frequently in meetings of men with curiousity, attracts the attention of many with regard to the opinion of Copernicus which he holds to be true ... he talks frequently with fifteen or twenty guests who argue with him now in one house, now in another. But he is so well fortified that he laughs them off; and although people are left unpersuaded because of the novelty of his opinion, still he shows up as worthless the majority of the arguments with which his opponents try to defeat him. Monday, in particular, in the home of Federico Ghisilieri, he was especially effective. What I enjoyed most was that before he would answer the arguments of his opponents, he would amplify them and strengthen them with new grounds which made them appear invincible, so that, when he proceeded to demolish them, he made his opponents look all the more ridiculous.[1]

His *modus procedendi* in the circumstances was incredibly bold. In his defense it may be said that he felt certain that such a procedure would draw attention to his arguments and perhaps their force would be enough to carry the day. Even if Galileo was justified in rejecting the compromise in that he wanted to hypothesize about the real world and not merely formulate an abstract mathematical design which saved the appearances, he was not justified in claiming superior proof for his system over any other until he had such proof. It is true that he was a cham-

---

[1] Letter of Antonio Querengo to Alessandro D'Este, dated January 20, 1616. *Opere*, XII, 226 f.

pion of the new experimental method and as such, was one of the few who felt that the new science needed independence from outside influences if it was to prosper. But by claiming more strength for his observations than they really had, he was defeating his own purpose. Since this was the first major conflict of the new science with the old, it might have been better had Galileo proceeded more cautiously and made sure that he could back up his claims with evidence which would compel their acceptance and with the new physical doctrines which the Copernican theory, taken as a reality, necessarily implied.

Galileo's activities in Rome must have been a great disappointment to the Jesuit astronomers there. They had acknowledged his discoveries and the consequences which followed from them. It is true that they had to be cautious for a number of reasons, not all of them scientific. But they were well aware, and many of them admitted it in so many words, that Ptolemy's system in *forma pura* was no longer tenable. Shortly before his death in 1612, the great Father Clavius had admitted as much. But there was still the matter of Tycho Brahe's alternative system. Unless they were able to get more proof for the Copernican theory, they would never be able to rule out the possibility of a geostatic universe.[2] That proof would have to come in physics and astronomy. Now the issue was being forced and a premature decision was imminent.

The Dominicans too were far from pleased with Galileo's conduct. It was their right and duty to be preeminently the defenders of the Faith. They must have seen Galileo with his private interpretation of Scripture and his demand for radical changes in the world-view as constituting a threat to the faith of the masses.

As we have seen, there were many reasons why theologians opposed the Copernican system. Besides the desire to "protect" Scripture and to guard man's position as the focal point of ma-

---

[2] The Jesuits of the Roman College had adopted the Tychonic system as a step in the direction of Copernicanism. They knew as well as anyone that unqualified support of a heliocentric system was impossible until Tycho's compromise was ruled out.

terial creation, there was a general feeling of siege brought about by the Reformation and a rigidity of the "official" philosophy and theology since they had ceased to be creative.

Still, there were cardinals, Jesuits, and Dominicans willing to come out in favor of Galileo. His most eloquent support came from the Dominican friar, Thomas Campanella. Campanella, one of the most striking figures of the age, met Galileo at Padua in 1592. Their friendship seems to have been partially based on a common dislike of Aristotelian philosophy, which Campanella felt was pagan and dogmatic. His brilliant but undisciplined mind was captivated by the naturalistic philosophy of Bernardino Telesio. With Galileo, he advocated a direct study of nature and not just ancient texts about nature. But while Galileo was a constructor of the new science, Campanella was only its exaltor, its cheering prophet.

In August, 1599, Campanella was implicated in a conspiracy which plotted to overthrow the Spanish government of Naples. On September 6 he was arrested in Calabria and brought to Naples for trial. Charges of heresy and political rebellion were brought against him so that his trial was simultaneously ecclesiastical and secular.[3] It was claimed that he was mixed up with magic and that he had predicted the fall of the Spanish government by astrology. The Spanish government wanted him sentenced to death for treason, but he was sentenced instead to life imprisonment in the fortress of Naples.

Campanella continued to study and write while in prison. A number of his manuscripts were entrusted to Tobias Adam and published in Germany. He was allowed to correspond with the leading intellectuals of his time, and in this way kept himself informed with regard to currents of thought in science, philosophy, and theology. He understood the main conclusion of Galileo's discoveries in 1610: Aristotelian cosmology was beginning to topple. He wrote to Galileo in 1611 and encouraged him not to be like Columbus, who, after he had discovered a new world,

---

[3] See P. Mandonnet, art. "Campanella," *Dictionnaire de théologie catholique* (Paris: 1932), II, 2, 1443.

left it to others to explore and conquer. But he cautioned, "Many Copernican doctrines must be changed by you." [4]

While Galileo was campaigning in Rome in 1615, Boniface Cardinal Gaetani wrote to the imprisoned Dominican and asked his opinion on the whole matter. This is another indication that there were important people in Rome who considered the issue to be more than an open and shut case against Galileo.

Campanella's answer to the Cardinal was a short treatise entitled *Apologia pro Galileo Mathematico Florentino*. The *Apologia* did not analyze the scientific arguments. It concentrated on the question of whether or not Galileo was free to construct his new science or whether he and his doctrine should be condemned in the name of patristic tradition and Holy Scripture.

In the first of five chapters in his treatise, Campanella listed the arguments used by philosophers and theologians against Galileo. The Florentine scientist had, so the objections said, contradicted Aristotelian philosophy and the unanimous opinion of the Fathers, misinterpreted Sacred Scripture, postulated a plurality of worlds, and invented novelties. In the second chapter of his work, Campanella noted the arguments which supported Galileo. He pointed out that nearly seventy years had passed since Copernicus had dedicated his work, without objection, to Pope Paul III. Since that time, great men such as Reinhold, Maestlin, Kepler, Gilbert, and Magini had accepted the Copernican theory.

> Galileo has made no additions to the system advanced by Copernicus, and if the *De revolutionibus* does not adversely affect Catholic faith, neither does the work of Galileo. [5]

Another argument for Galileo was that:

> Because heaven is immobile, Holy Scripture names it the Firmament. The earth therefore must rotate, and the sun stand in the center of the world. As Copernicus and his followers prove and

[4] *Opere*, XII, 21.

[5] *Apologia pro Galileo* (Frankfort: 1622), p. 9. I have used Grant McColley's translation (Smith College Studies in History, XXII, n. 3–4: April-July, 1937) with some modifications. All references to the Apologia are to the original Latin text. I hope to publish soon an accurate translation of, and commentary on, this important work.

the followers of Ptolemy now admit, this system explains all phenomena and accords with all principles of mathematics.[6]

It must be noted that Campanella was merely reporting the arguments in favor of Galileo. He himself was not convinced that the Copernican theory was true. Because he wanted to see a Christian philosophy develop and flourish, the Dominican was especially interested in advocating Galileo's right to speculate. He wrote:

> Because knowledge should be Christian, they lack understanding who forbid and prohibit philosophy among the followers of Christ. They are like the Emperor Julian, who interdicted and outlawed from the faith all the sciences of the Christians.[7]

Chapter three postulated certain indispensable prerequisites for arriving at a solution of the problem. For anyone to be a judge in this matter, he must not only know Holy Scripture and the Fathers, but science as well. The judges should bear in mind that:

> ... not all that is false is injurious to Scripture. It must directly or indirectly destroy the true meaning. In addition, if a theologian has advanced doctrines which apparently are equally or more opposed to the Scripture than are the theories of Galileo, he is neither condemned nor prohibited from making further inquiry. It is by such inquiry that he determines whether the doctrines advanced are sound. He does not impugn faith, but rather, opens truth to the soul.[8]

Campanella did not believe that any philosopher or theologian had yet formulated a wholly satisfactory system of the universe so that:

> It is an essential part of the glory of the Christian religion that we permit [Galileo's] method of discovering new knowledge and of rectifying the old.[9]

Next he makes it clear that no one has the right to make his own unofficial interpretation of Scripture as authoritative as Scripture itself:

[6] *Ibid.*, p. 11.
[7] *Ibid.*, pp. 25 f.
[8] *Ibid.*, p. 14.
[9] *Ibid.*, p. 26.

I have shown that liberty of thought is more vigorous in Christian countries than it is in other nations. Now if this be true, whoever on his own proscribes limits and laws for human thought and thinks his action is in harmony with the dictates of Holy Scripture, he is not only irrational and dangerous, but irreligious and impious as well. And I say the same of anyone who teaches and accepts no interpretation but his own, and subjects Scripture to his beliefs or to those of some other writer. Such a practice exposes Holy Scripture to the jest of unbelievers and heretics.[10]

Using St. Thomas Aquinas and St. Augustine as his authorities, he warns against too literalist an interpretation of the texts in question:

In the preface to his *Tract against the Errors of the Greeks*, St. Thomas states: 'I first assert that many Scriptural passages do not pertain to dogmas of faith, but rather to doctrines of philosophy. It does great violence to such passages to affirm or deny them as if they are dogmas.' . . . St. Augustine expresses a similar view in the first chapter of his *Commentary on Genesis*: 'It is greatly to be guarded against, terrible and shameful for a Christian to speak of physical phenomena as if he were discussing Scripture.' [11]

And, Campanella cautions,

If Galileo has demonstrated conclusively the things which he asserts, there will insue among heretics no slight mockery of our Roman theology . . . If Galileo's hypothesis is false, it will not disturb theological doctrine, for not everything that is untrue is contrary to faith. If everything false were contrary to faith, the errors we find made by some saints in treating of natural philosophy would make them heretics. If Galileo's theory is unsound, it will not endure. I think therefore that his type of philosophy should not be forbidden.[12]

In chapter four, Campanella replied to the objections listed against Galileo in the first chapter of the treatise. His main point was that Scripture and the Fathers do not supply sufficient evidence on which to construct or prohibit any cosmological system. The final chapter was an examination of the arguments for Gali-

[10] *Ibid.*, p. 27.
[11] *Ibid.*
[12] *Ibid.*, p. 29.

leo. Campanella admitted that he himself had tried, in some of his writings, to refute the arguments of Copernicus and Pythagoras. But since Galileo's discoveries, he had had to change a number of his opinions. He still had reservations with regard to the new astronomy: "The hypothesis of Copernicus and Galileo is probable, not true [i.e., not demonstrated]." [13] Campanella considered the Copernican system to be a solid theory, but he did not come out in unqualified support of it. He concluded the *Apologia*:

> It is unnecessary that the investigations of Galileo be suppressed, a misfortune which is about to occur. Our enemies will seize eagerly upon this action and proclaim it abroad.[14]

The *Apologia* was not without logical flaws and it showed a subjective prejudice against everything Aristotelian as well as several theological wanderings. But, on the whole, it was an amazing document. Theologically more correct than Bellarmine's *Letter to Foscarini* and, in a way, more honest than Galileo's *Letter to the Grand Duchess*, it was the best evaluation of the state of affairs yet formulated. But Campanella was not a popular figure in Rome and his *Apologia* passed unnoticed, except by a few, in the turmoil of the Eternal City.[15]

In Rome, Galileo's campaign continued. Early in 1616, Guicciardini reported to the Grand Duke:

> He is passionately involved in this fight of his, and he does not see or sense what it involves, with the result that he will be tripped up and will get himself into trouble, together with anyone who supports his views. For he is vehement and stubborn

[13] *Ibid.*, p. 54.
[14] *Ibid.*, pp. 57 f.
[15] As to Campanella, in 1626, Pope Urban VIII ordered Naples to send him to Rome, ostensibly for punishment. In reality, it was a device to outwit the Spaniards. Campanella gained his complete freedom in Rome on April 6, 1629. The Dominican friar remained in Rome until 1634, when another Calabrian plot was discovered, and, since it had been engineered by one of Campanella's disciples, the Pope advised him to escape to France to avoid another imprisonment at the hands of the Spanish. He was welcomed to France by Cardinal Richelieu who granted him a yearly pension. Campanella died at the Dominican convent of St. Honoré in Paris on May 21, 1639.

and very worked up in this matter and it is impossible, when he is around, to escape from his hands. And this business is not a joke, but may become of great consequence, and the man is here under our protection and responsibility.[16]

Galileo was able to write on February 6, 1616, that the gossip about his being a heretic and blasphemer had been quieted. This was due, in large measure, to the fact that the Holy Office had judged the *Letter to Castelli* and Caccini's testimony, in favor of Galileo. The day before he wrote that his reputation was cleared, February 5, Galileo was visited by Tommaso Caccini. While the exact purpose of this visit is unknown, de Santillana speculates that he was told by the Holy Office to apologize to Galileo for the charges he had made against him and to find out if Galileo would agree to withdraw from the controversy.[17] This is entirely possible. At any rate, Galileo did not withdraw. Instead he tried to convince the Pope himself that the new astronomy had to be accepted. He secured a letter from the Grand Duke recommending him to Cardinal Orsini. He explained to Orsini that he now had a conclusive proof for the heliocentric system.[18] Would the Cardinal present the new argument to Pope Paul V for him? Orsini agreed.

Judging from a letter written by Guicciardini to the Grand Duke, Orsini's urging of Galileo's new proof was the catalyst needed to convince the Pope that this whole matter had gone far enough.[19] On February 19, 1616, the theological consultors of the

[16] *Opere*, XII, 243.

[17] G. de Santillana, *Crime of Galileo*, p. 121, n. 5.

[18] This was his theory of the tides, an attempt to prove that the earth's double motion is responsible for the flux and reflux of the sea. Kepler's view, which was well known even in the seventeenth century, was the correct one, namely, that the tides are caused by the attraction of the moon. For an excellent modern consideration of the tides, see Albert Defant, *Ebb and Flow* (Ann Arbor: University of Michigan Press, 1958).

[19] It has often been said that Pope Paul V was an anti-intellectual. This is largely because Guicciardini described him as a man who abhorred the liberal arts and could not stand novelties and subtleties. But there might have been good reasons for his flinching from subtle arguments. He had assisted in person at the long, involved disputations of the *Congregatio de Auxiliis* on the question of grace and free will. This dispute, mainly between the Dominicans and Jesuits, began in 1599 under Pope Clement VIII and dragged on, without solution, into the papacy of Paul V. Pope Paul finally

Holy Office were summoned to give a formal decision on the Copernican system.

The Sacred Congregation of the Holy Office was composed then, as now, of a number of cardinals who have as their advisors certain religious and secular clerics, learned in Canon Law, Conciliar Law, and Sacred Theology and appointed by the pope to their posts as *Consultores* of the Holy Office. According to John Baptist Cardinal de Luca, a junior contemporary of Galileo, the general procedure followed in such matters was this:

> A preparatory meeting is held on Saturdays to decide what business should go to the Consultors (or Qualifiers, as they were also called, since part of their duty was to decide what particular note of condemnation was to be attached to propositions which they judged to be unorthodox [20]) and what should be handled directly by the cardinals themselves. This meeting is attended by only six officials of the Holy Office. On Mondays, the Consultors meet and give their opinion on the subjects to be presented to the cardinals. On Wednesdays, the meetings of the cardinals are held. The opinions of the Consultors are heard and discussed and the cardinals then give their opinion. The following day, Thursday, some of the cardinals meet in the presence of the Pope, for whom the matter is summarized, the vote of the Congregation revealed, and the matter resolved.[21]

---

ended the discussion in 1607 by ordering the Dominicans and Jesuits to refrain from attacking each other's position as heretical. The Galileo affair must have seemed to the Pope to be a repetition, on a lesser scale, of the controversy on grace, and, remembering all the unrest the *De Auxiliis* had caused, he decided to resolve this problem before it grew to the status of a full-blown battle. This call for a verdict, because it was hasty and premature, was certainly an error.

[20] Lest there be any confusion with regard to terminology, it may be noted that Church law made provision for Qualifiers who were to assist the Consultors by assigning the particular censure which an unorthodox proposition deserved. But a Canon Lawyer of the seventeenth century notes that in Rome, the Consultors often assumed the duties of the Qualifiers. This was true in the Galileo decision since we are certain that at least six of those who censured the two propositions were Consultors of the Holy Office. Thus I prefer to consider the decision as coming from Consultors rather than Qualifiers, though the latter title is also valid. See P. Passerini, *Commentarium in quartum et quintum librum sexti decretalium* (Rome: 1674), III, p. 220.

[21] John Baptist Cardinal de Luca, (d. 1683), *Theatrum veritatis et justitiate* (Venice: 1698), XV, pars II, disc. XIV, pp. 50 f.

Two propositions representing Galileo's doctrine were submitted to the Consultors for their opinion.

> I. The sun is the center of the world and completely immovable by local motion.
> II. The earth is not the center of the world, nor immovable, but moves according to the whole of itself, and also with a diurnal motion.[22]

The Consultors met on February 23 and decided upon the following censures:

> The first proposition was declared unanimously to be foolish and absurd in philosophy and formally heretical inasmuch as it expressly contradicts the doctrine of Holy Scripture in many passages, both in their literal meaning and according to the general interpretation of the Fathers and Doctors.

With regard to the second proposition,

> All were agreed that this proposition merits the same censure in philosophy, and that, from a theological standpoint, it is at least erroneous in the faith.[23]

The theologian Antonio of Cordova, writing in 1604, explains the generic meaning of these censures.[24] The "formally heretical" in the first censure means that this proposition was considered directly contrary to a doctrine of faith. This shows that the apparent affirmations of Scripture and the Fathers that the sun moves, was held by the Consultors to be a doctrine of faith. In other

---

[22] "*Prima: Sol est centrum mundi, et omnino immobilis motu locali.* 2a: *Terra non est centrum mundi nec immobilis, sed secundum se totam movetur, etiam motu diurno.*" *Opere,* XIX, 321. The strange wording of these propositions seems to be an attempt to translate the Copernican position into some kind of formal philosophical language.

[23] "*Omnes dixerunt, dictam propositionem esse stultam et absurdam in philosophia, et formaliter hereticam, quatenus contradicit expresse sententiis Sacrae Scripturae in multis locis secundum proprietatem verborum et secundum communem expositionem et sensum Sanctorum Patrum et theologorum doctorum.*"
2a. "*Omnes dixerunt hanc propositionem recipere eandem censuram in philosophia; et spectando veritatem theologicam, ad minus in Fide erroneam.*" *Opere,* XIX, 321.

[24] Antonius Cordubensis, *Quaestiones Theologicae* (Venice: 1604), I, q. 17, 146 ff. Also see John Cahill, *The Development of Theological Censures after the Council of Trent* (1563–1709), (Fribourg: University Press), pp. 174 ff.

words, there is no room for apologetic excursions here. The Consultors tagged the proposition with the strongest possible censure, as being directly contrary to the truth of Sacred Scripture. In the second proposition, the motion of the earth was censured as "erroneous in the faith." This meant that the Consultors considered it to be not directly contrary to Scripture, but opposed to a doctrine which pertained to the faith according to the common consensus of learned theologians. In other words, Scripture was not as definite in stating the immobility of the earth. But the Holy Writ did "reveal" that the sun moved, and since human reason could conclude that the sun and the earth were not both moving around each other, the Consultors felt that the immobility of the earth was a matter which fell under the domain of faith indirectly, as a kind of theological conclusion.

Both opinions were completely wrong. No Church historian can defend them. But it is legitimate to point out that the error had a number of causes, which, though they do not justify it, do help to explain how it could happen. Mainly the blame must fall on the eleven Consultors. They were not mere rubber stamps. Most of them were very eminent theologians who were far from afraid to speak out in defense of their opinions. Thomas de Lemos, Peter Lombard, and Gregory Coronel had all been heatedly involved in the *Congregatio de Auxiliis*. But since there was not a scientist among them—they were theologians, not mathematicians or astronomers—they were out of their field. But how, even if they knew no science, could they be so wrong in theology? That was their field and their fault. Here a number of factors must be given a share of the blame. Theological circumstances fostered their radically conservative position. The Consultors were sensitive with regard to anything which required or anyone who demanded a novel interpretation of Scripture. But it remains something of a mystery why, instead of paying attention to Augustine, Aquinas, Campanella or even Galileo, the Consultors adopted as a matter of course Cardinal Bellarmine's faulty exegetical procedure. This was nobody's fault but their own. They are the ones who actually judged that the new astronomy contradicted Sacred Scripture. Pope Paul V is also partly responsible for the error. He had called for a hasty and premature decision. Galileo is not completely blameless. Brodrick is right:

> He believed completely in the physical truth of Copernicanism and was certain that it would eventually be recognized by all mankind. But he wanted it recognized almost overnight, and thereby hangs the miserable and entirely unnecessary tale of his condemnation.[25]

Giorgio de Santillana claims that Galileo's lack of proof was not an important factor in the Consultors' decision.[26] But this simply is not true. Had there been more proof, I think the qualifications would have been considerably less severe. Professor de Santillana's argument is that Campanella's *Apologia* does not list insufficient proof as one of the arguments against Galileo. Perhaps de Santillana could profit from a closer study of Campanella's *Apologia*. It was a *defense* of Galileo. Had Campanella listed lack of proof in chapter one as an argument against Galileo, he would have had to respond to it in chapter four where he answered those objections. He knew that his defense of Galileo would suffer if he brought in the matter of scientific proof. He personally held that "the hypothesis of Copernicus and Galileo is probable, not true." [27] In other words, Campanella was not defending the scientific side of Galileo's case, he was defending the scientist's right to theorize within a just system of authority. Everyone knew that Bellarmine had challenged Galileo to come forth with solid proof. The Jesuit astronomers of the Roman College were quite certain that Galileo had gone beyond his evidence in asserting Copernicanism to be a fact. Father Grienberger, the leading Jesuit astronomer, said frankly that Galileo would do better to produce more convincing proofs for his theory before trying to adjust Scripture to fit it.[28]

[25] J. Brodrick, *Robert Bellarmine: Saint and Scholar*, p. 368.
[26] G. de Santillana, *Crime of Galileo*, p. 143, n. 19.
[27] Campanella, *Apologia pro Galileo*, p. 54.
[28] *Opere*, XII, 151.

On February 25, the day after the Consultors turned in their decisions,

> The Lord Cardinal Mellini notified the Reverend Fathers, the Assessor, and the Commissary of the Holy Office that the censure passed by the theologians with regard to the propositions of the mathematician Galileo, to the effect that the sun is the center of the world and immovable by local motion, and that the earth moves, and even with a diurnal motion, had been reported; and His Holiness had directed Cardinal Bellarmine to summon before him the said Galileo and admonish him to abandon the said opinion; and, if he should refuse to obey, the Commissary should enjoin on him, before a notary and witnesses, a command to abstain altogether from teaching or defending this opinion and doctrine and even from discussing it; and if he does not acquiesce therein, he should be imprisoned.[29]

For the next day, February 26, the following report can be found in the files of the Holy Office:

> At the Palace, the usual residence of the aforenamed Cardinal Bellarmine, the said Galileo, having been summoned and standing before His Lordship, was, in the presence of the Very Reverend Father Michael Angelo Seghiti de Lauda, of the Order of Preachers, Commissary-General of the Holy Office, admonished by the Cardinal of the error of the aforesaid opinion and that he should abandon it; and later on—*successive ac incontinenti*—in the presence of myself, other witnesses, and the Lord Cardinal, who was still present, the said Commissary did enjoin upon the said Galileo, there present, and did order him (in his own name), the name of His Holiness the Pope, and that of the Congregation of the Holy Office, to relinquish altogether the said opinion, namely, that the sun is in the center of the universe and immobile, and that the earth moves; nor henceforth to hold, teach, or defend it in any way, either verbally or in writing. Otherwise proceedings would be taken against him by the Holy Office. The said Galileo acquiesced in this ruling and promised to obey it.
>
> Done at Rome, in the above-mentioned place, in the presence of the Reverend Baldino Nores from Nicosia in the Kingdom of Cyprus, and Augustino Mongardo, of the diocese of Montepulciano, both witnesses belonging to the household of the said Lord Cardinal.[30]

29 *Opere*, XIX, 321.
30 *Opere*, XIX, 321 f.

This document has been the subject of a long and still unresolved debate. A number of historians have tried to show that the report of February 26 is spurious and that it was inserted into the file as early as 1616 or as late as 1632 in order to ensnare Galileo in a preconceived trap.

There is no doubt that the report is defective in form. The instruction from the Pope was that should Galileo refuse to obey, the Commissary was to "enjoin upon him before a notary and witnesses, a command to abstain altogether from teaching or defending this opinion and doctrine and even from discussing it." First of all, then, the Commissary was not to serve an injunction upon Galileo unless he refused to accept Cardinal Bellarmine's admonition. But, in the report of February 26 there is no mention of Galileo objecting to Bellarmine's warning. Secondly, the record says that "Cardinal Bellarmine admonished Galileo to abandon the Copernican theory, and then (*successive ac incontinenti*, which de Santillana and most others translate "immediately thereafter"), the Commissary served an absolute injunction, without, it appears, giving Galileo time either to acquiesce in or to refuse the Cardinal's admonition. Thirdly the instruction specifies that a notary and witnesses should be present as witnesses of the Commissary's action presumably for the purpose of signing the official account later. Yet the report of February 26 does not carry the signature of the notary, and the witnesses are from Cardinal Bellarmine's personal staff and are not apparently personnel of the Holy Office. Thus they would have no business witnessing the official acts of the tribunal. Fourthly, the document of February 26 is only a *registratur* or report, which explains why it is unsigned, but means that the official protocol which should have been made out at Bellarmine's residence and inserted in the Holy Office files is missing. In other words—the record we have is not the official, signed report one would expect to find, but an administrative minute entered in the Acts of the Holy Office that seems to be referring to an official document which is not there.

Emil Wohlwill, in 1870, attempted to show that the report of the audience with Bellarmine and the Commissary had been tampered with.[31] After studying the original pages, he concluded that when the Galileo matter came up before the Holy Office again in 1632, someone in the Holy Office who had access to the Galileo file erased the words "the said Galileo acquiesced in this ruling and promised to obey it" which, he says, came right after the Cardinal's admonition. He then added on a completely false account of what happened and this is what we have in the files today. Thus instead of reading "admonished by the Cardinal of the error of the aforesaid opinion and that he should abandon it, Galileo acquiesced in this ruling and promised to obey it," the "forged" account reads "...admonished by the Cardinal of the error of the aforesaid opinion and that he should abandon it; and immediately thereafter (*successive ac incotinenti*:), in the presence of myself," etc.

Against Wohlwill's thesis, however, stands the careful research of Laemmel, who, after examining the record with X-ray and ultraviolet tests, proved beyond doubt that there had been no erasing or writing in on the folio pages in question.[32] Furthermore experts certified that the Galileo record was written by the same hand as the preceding and following documents in the file which were definitely written in 1616. Thus there was no forgery involved.

Karl Von Gebler and, more recently, Giorgio de Santillana contend that the report of an absolute injunction served upon Galileo by the Commissary-General was inserted in the file in 1616 by one or more officials of the Holy Office just in case Galileo would decide to act up again.[33] Thus, de Santillana writes:

> We may then reconstruct as follows: the Commissary, as he watched the scene (we know he was present), was disgusted with the easy way Galileo was let off, and he decided to omit

[31] Emil Wohlwill in his *Der Inquisitionsprozess des Galileo Galilei* (Hamburg: 1870), was the first to question the validity of the strict injunction.

[32] See H. Laemmel, *Archiv f. Gesch. d. Mathematik*, X, (March, 1928), cited by de Santillana, *op. cit.*, p. 264.

[33] See Karl Von Gebler, *Galileo and the Roman Curia* (London: 1879), esp. 89 and 322 ff.

the protocol, although his instructions were clear, and the witnesses already designated, obviously by the Cardinal himself. On going back to his office, he told his assistant to arrange a more helpful minute of the proceedings. 'And,' he may have added, 'make it stiff, just in case. What they don't know doesn't hurt them; when trouble arises, it is we who have to take it on.' [34]

Von Gebler, de Santillana, and those who share their view also point to a report given on March 3, 1616, at the first meeting of the Holy Office since Galileo had been summoned to appear before Cardinal Bellarmine. It reads:

> The Lord Cardinal Bellarmine having reported that Galileo Galilei, mathematician, had, in terms of the order of this Holy Congregation, been admonished to abandon the opinion which he has held up to that time, that the sun is the center of the spheres and immobile and that the earth moves, and had acquiesced therein . . .[35]

The report mentions no command of the Commissary-General to Galileo. Still the opinion that the Commissary, either on his own or in collusion with others, had a completely false story inserted into the official records in order to trip up Galileo, if that became necessary, seems improbable for two reasons. First, the supposed malefactors would have needed the gift of prophecy and a great deal of luck to forsee that Galileo would be condemned seventeen years later partly on the basis of this document, and, they could only hope that no one would recognize it as spurious. Secondly, a planted document would be of no value as long as Cardinal Bellarmine and the other officials present at the meeting with Galileo were still alive. They certainly would have remembered that no such injunction had been given and would have dealt severely with whoever had falsified the official records of the Holy Office. This was serious business and it was commonly held by moral theologians that any such tampering was seriously sinful and worthy of grave penalties.

Still more confusion concerning the absolute injunction arises from the use of the words *"successive ac incontinenti"* in the

[34] Giorgio de Santillana, *op. cit.*, p. 266.
[35] *Opere*, XIX, 336.

report. Giorgio de Santillana and others translate them as "immediately thereafter" which would mean, as we have pointed out, that Galileo never had a chance to acquiesce in Bellarmine's admonition and any absolute command would have been totally out of order. F. H. Reusch has shown that the words *"successive oc incontinenti"* in the Vatican usage of that time do not mean "immediately thereafter" or "without pause" but "in the sequel" or "later on." [36] This would allow for the possibility that Galileo tried to defend his theory after Bellarmine's warning and then the Commissary-General stepped in with the absolute injunction.

Father Joseph Clark, S.J., offers a third possibility. As Father Clark translates the important passage it reads:

> ... the same illustrious Lord Cardinal warned the above-mentioned Galileo about the error of his opinion described above, and advised him to abandon it, and over, and over and over again, and since the aforesaid Galileo was uncontrollably voluble on the matter, before me and other witnesses, etc.[37]

The confusion arises from the fact that none of these three translations nor the important implications following upon which one is accepted, can be disproved on the basis of the text itself.

What, then, are we to conclude with regard to this important document? There are good arguments pro and con concerning its validity as an authentic report. The evidence at hand simply does not afford certitude one way or the other. Perhaps more documents will turn up eventually which will throw more light on this particular aspect of the case. But until such evidence is forthcoming, it would seem a bit tenuous to build a historical account of cloak and dagger treachery around such a disputed document, especially when the document can be given a perfectly sound historical interpretation without a trace of falsehood or deception.

Until recently, I was of the opinion that Galileo, who, after

[36] See F. H. Reusch, *Der Process Galilei's und die Jesuiten* (Bonn: 1879), pp. 137 ff.

[37] Fr. Joseph Clark, S.J., in a talk given on April 11, 1964, at the Galileo Centennial held at the University of Notre Dame.

all, was not one to sit quietly and listen, began to defend himself after hearing the Cardinal's admonition, not knowing that the consequence would be an absolute injunction. The injunction would then fall outside of any "plot" theory and might not even bind Galileo because of his failure, in the circumstances, to understand exactly what was going on.

I am now inclined, however, to accept the position of Stillman Drake, who holds that after Bellarmine's warning and before Galileo could agree to it, the Commissary-General, overzealous in the fulfillment of his assignment, stepped forward and issued the strict injunction to which Galileo, who must have been somewhat confused by this action, simply agreed. Later, Bellarmine told Galileo that the Commissary-General's action was illegal and not in accord with his instructions and therefore he, Galileo, was not bound to follow it, and, in fact, should forget that it happened. This seems to be supported by the confidence with which Galileo asked Bellarmine for a certificate in writing with which to defend himself from the rumors that he had been forced to abjure his opinion. But more on that later.

In any event, with Galileo's promise to comply with the ruling, the case was closed so far as the Holy Office was concerned. Next, the decision of the Holy Office was referred to the Congregation of the Index for action. Until the year 1571, the Holy Office had, in addition to its other duties, been in charge of the censorship of books. When that task became too time-consuming, a separate Congregation was created and given the responsibility of judging whether books were or were not fit reading for the faithful. Pope Pius V founded the Congregation of the Index in 1571. It was suppressed on March 25, 1917, by Pope Benedict XV, with the censorship of books reverting to the Congregation of the Holy Office.

On March 5, the Congregation of the Index published a decree regarding the Copernican astronomy:

> ...And because it has also come to the attention of this Congregation that the Pythagorean doctrine, which is false and

contrary to Holy Scripture, which teaches the motion of the earth and the immobility of the sun, and which is taught by Nicholas Copernicus in *De revolutionibus orbium caelestium* and by Diego de Zuñiga's *On Job*, is now being spread and accepted by many—as may be seen from a letter of a Carmelite Father, entitled, *Letter of the Rev. Father Paolo Antonio Foscarini Carmelite, on the Opinion of the Pythagoreans and of Copernicus concerning the Motion of the Earth and the Stability of the Sun, and the New Pythagorean System of the World*, printed in Naples by Lazzaro Scoriggio in 1615: in which the said Father tries to show that the doctrine of the immobility of the sun in the center of the world, and that of the earth's motion, is consonant with the truth and is not opposed to Holy Scripture. Therefore, so that this opinion may not spread any further to the prejudice of Catholic truth, it [the Sacred Congregation] decrees that the said Nicholas Copernicus, *De revolutionibus orbium*, and Diego de Zuñiga, *On Job*, be suspended until corrected; but that the book of the Carmelite Father, Paolo Antonio Foscarini be prohibited and condemned, and that all other books likewise, in which the same is taught, be prohibited, as this decree prohibits, condemns, and suspends them all respectively. In witness thereof, the present decree has been signed and sealed by the hand and seal of the Most Eminent and Reverend Lord Cardinal of St. Cecilia, Bishop of Albano, on the fifth day of March, 1616.[38]

It is important to note that while Copernicus's *De revolutionibus* and de Zuñiga's *On Job* (a commentary published in 1584 which discussed the Copernican theory and the possibility that it was not against Holy Scripture), were suspended until corrected, Foscarini's book was condemned outright. This was an application of the distinction between hypothesis and fact. Copernicus's work presented a hypothetical astronomy which, if considered as a mathematical device only, did not contradict Scripture. De Zuñiga had discussed the compatibility of the Copernican theory with Scripture in one place only and that could easily be deleted. But Foscarini had attempted to demonstrate that the Copernican doctrine was "consonant with truth and not opposed to Holy Scripture." The attitude expressed in the decree seems to have been that Copernicus's system taken as a scientific hypothesis was harmless

[38] *Opere*, XIX, 323.

enough; but let there be no more attempts to demonstrate that this system is true in fact or, especially that it is in conformity with Sacred Scripture.[39] The Copernican system though "false and contrary to Scripture" could be discussed in terms of hypothesis but not as fact. It could not be discussed at all as a system possibly reconcilable with Scripture.

The decree raises another issue for discussion, that of the Catholic doctrine of infallibility. No serious scholar today holds that the Church intended to make an infallible commitment against the Copernican astronomy. But as long as mistaken and misguided attacks such as that of A. D. White [40] are current, clarification is necessary to prevent their acceptance on faith by students and casual readers. The fact is that White's charges, as well as the innuendos frequently found in popular history books, are made without any knowledge of the Roman Catholic Church's actual teaching on infallibility.[41]

[39] As Galileo interpreted the decree, it forbade "only those books which *ex professo* try to prove that [the Copernican system] is not contrary to Scripture." *Opere*, XII, 244. As Bellarmine interpreted it for him, it meant that this system was not to be defended or held.

[40] Andrew D. White, in his *History of the Warfare of Science with Theology in Christendom* (New York: Dover, 1960), I, pp. 158–170, completely misrepresents the facts and shows a total lack of understanding regarding the theological doctrines he tries so hard to destroy.

[41] Infallibility is difficult to explain in a brief statement. Basically, it means that Christ bestowed divine assistance in the form of infallible guidance upon His Church. This is the infallibility which was conferred on Peter and his successors and which was given for the good of the whole Church. One and the same infallibility is found in the pope and the Church. The infallibility of the Church is "not created by the pope, but strengthened by him; he is united to it as its head." (Address of Archbishop Descuffi to the Second Vatican Council in *Council Speeches of Vatican II* [New York: Deus Books, 1964], p. 70.) This infallibility is active (the Teaching Church) and passive (the Learning Church). The pope can exercise it even without the consent of the Learning Church. Also, the universal consent of the Learning Church with regard to a matter of faith or morals is infallible. In addition, the First Vatican Council teaches, "One must believe by divine and Catholic faith all things contained in the word of God, whether written or traditional and which are proposed as divinely revealed for belief by the Church, either by a solemn judgment or by the ordinary and universal magisterium" (Session III, chapter 3). The ordinary and universal magisterium is evidenced in the unanimous preaching of the bishops of the world in union with the pope. The extraordinary magisterium is exercised by the pope speaking *ex cathedra* and by the

Briefly, the Roman Catholic Church teaches that the pope is infallible when he pronounces *ex cathedra* on a matter of faith or morals.[42] This means a) that in making such pronouncements, the pope must be speaking to the entire Church, b) that he obviously acts in virtue of his full authority, c) that he manifestly wishes to define a doctrine in an irreformable manner, and, d) that the context of the proposition defined is matter dealing with faith or morals. Obviously then, not every official act of a pope is infallible. This is not a new doctrine invented by apologists. It was very clear in Galileo's time. Cardinal Bellarmine himself wrote that:

> We do not deny that popes are able, by their example, to present the occasion of erring. But we do deny this, that they can prescribe *ex cathedra* some error which must be followed by the whole Church.[43]

The great Dominican theologian Dominic Bañez (1528–1604) admitted long before the Galileo affair that the "pope as a private person can err even in matters of faith and morals, but not when he defines a doctrine as the pope of the Church." [44] Nor can the pope delegate his personal prerogative of infallible authority to any Congregation or individual. Bañez explicitly states that "the pope cannot delegate the power of defining infallibility in matters of faith." [45] It is true that the Roman Congregations share in the government of the Church and thus in the pope's

---

Ecumenical Councils with the approval of the pope. The ordinary magisterium as exercised in papal encyclicals demands true internal assent of believers, but it is debated among theologians when and if such teachings enjoy infallibility. See Charles Journet, *The Church of the Word Incarnate* (New York: Sheed and Ward, 1954), I. The Church today seems to be coming to an even deeper realization of the meaning and extent of infallibility.

[42] Since the objections raised by the Galileo case are aimed specifically at the infallibility of the pope, we will confine our treatment of infallibility to the way in which it applies to papal decisions.

[43] St. Robert Bellarmine, *De potestate spirituali Summi Pontificis*, Bk. IV, ch. 8, 486. In *Opera Omnia Roberti Bellarmini*, ed. J. Guiliano (Naples: 1856), I.

[44] D. Bañez, *Commentaria in secundam secundae Sancti Thomae* (Venice: 1586), Q. 1, art. 10, 131e.

[45] *Ibid.*, 155d.

power of ruling, but they do not share in his dogmatic infallibility. The decrees of a Congregation may be doctrinal or disciplinary. And they may be approved by the pope in either of two ways: in *forma communi* (in a general way) or in *forma specifica* (in a special way). If they are approved in a general way, the decrees remain the acts of the Congregation from which they proceed, and they bind accordingly. Approval in a special way means that the pope adopts the decree of a Congregation and issues it in his own name. But even approval in this way does not imply commitment of infallible authority. As we have seen, the pope must proclaim dogmatic definitions *ex cathedra*. This requires a special act with very definite requirements.

With regard to the decree of the Congregation of the Index issued on March 5, 1616, we can say that it was both doctrinal and disciplinary: doctrinal in that it declared the Copernican system to be contrary to Holy Scripture, disciplinary because it suspended and prohibited the authors named in it. But the decree was issued in a reformable manner by a fallible authority. It was approved by the pope in the general way only (in *forma communi*) and thus remained an act of the Congregation of the Index, without any special endorsement of the Holy Father. Yet Professor White wrote that:

> The condemnations were inscribed on the Index and, finally, the papacy committed itself as an infallible judge and teacher to the world by prefixing to the Index the usual Papal Bull giving its monitions the most solemn papal sanction.[46]

White is referring here to the publication in 1664 of the *Index of Prohibited Books* which was prefaced, as is every new edition of the *Index*, by an Apostolic Letter which gave approval in a general way to the prohibitions published by the Congregation of the Index. It in no way involved special approval of the prohibitions, much less a commitment of infallible authority. Many books are placed on the *Index* and removed in subsequent edi-

---

[46] A. D. White, *op. cit.*, pp. 138 ff.

tions. The prohibition of books is not an irreformable act. But, as Koestler notes,

> It [the decree of March 5, 1616] was issued by the Congregation of the Index, but not confirmed by papal declaration ex *cathedra* or by an Ecumenical Council, and its contents therefore never became infallible dogma. All this was deliberate policy; it is even known that it was urged upon Paul V . . . by Cardinals Barberini and Gaetani. [This refers to a story, which might be true, that Barberini and Gaetani convinced the Pope that this was not a matter for dogmatic definition and that he should not commit papal authority against the Copernican theory.] These points have been stressed over and over again by Catholic apologists, but on the man in the street such subtleties were lost . . .[47]

Galileo remained in Rome for nearly three months after the case was closed. On March 11, he was received in audience by Pope Paul V and they talked for nearly an hour. The Pope assured Galileo that any rumors and calumny directed against him would be ignored by the Vatican. It did not take long for the rumors to begin. The publication of the decree of the Index gave Galileo's enemies ample material for malicious speculation about the theological orthodoxy of the Florentine scientist. A friend reported to Galileo that in Venice it was being said that he had been forced to abjure his opinion and had been severely censured by the Holy Office. Galileo appealed to Cardinal Bellarmine for a written statement which he could use in self defense. The Cardinal readily obliged him:

> We, Robert Cardinal Bellarmine, hearing that it has been calumniously rumored that Galileo Galilei has abjured in our hands and also has been given a salutary penance, and being requested to state the truth with regard to this, declare that this man Galileo has not abjured, either in our hands or in the hands of any other person here in Rome, or anywhere else as far as we know, any opinion or doctrine which he has held; nor has any salutary or any other kind of penance been given to him. Only the declaration made by the Holy Father and published by the Sacred Congregation of the Index has been revealed to him, which states that the doctrine of Copernicus, that the earth moves around the sun and that the sun is stationary in the center of

[47] Koestler, *The Sleepwalkers*, p. 457.

the universe and does not move from east to west, is contrary to Holy Scripture and therefore cannot be defended or held. In witness whereof we have written and signed this letter with our hand on this twenty-sixth day of May, 1616.[48]

Professor de Santillana concludes from the fact that no strict injunction is mentioned in Bellarmine's certificate, that none was given. But a close examination of the document reveals little support for this opinion. Theological writings dating from Galileo's time make it clear that to abjure an opinion meant to make a formal, external act of renouncing an offensive proposition, according to a fixed formula, and before ecclesiastical witnesses.[49] Thus it is true that Galileo had not been made to abjure, nor had he been given a penance to perform. But this does not mean that there was no strict injunction given to him "not to defend Copernicanism in any way, either verbally or in writing." I believe, with Stillman Drake, that such an injunction was given him, though illegally, by the Commissary-General, and it was Bellarmine who told him to ignore it. This written certificate was Galileo's defense should anyone ever try to trip him up on the basis of that injunction.

Getting back to the decree, there is no doubt that the prohibition of Copernicus's book until corrected was an unjust invasion of the rightful freedom of science. But it was not really a serious setback to progress in astronomy.[50] As a matter of fact, it was not until 1619, three years after the decree was issued, that John

[48] *Opere* XIX, 348.

[49] For a detailed consideration of abjuration, see J. Menochi, *De arbitrariis iudicum* (Venice: 1588), Bk. II, casus 372, "*Abiuratio haeresis quomodo sit faciendum.*"

[50] It might be asked what binding force the decree had on Catholic astronomers. First of all, they were obliged to conform exteriorly to the provisions of the decree. They could teach and write about the Copernican universe only as a hypothesis. Secondly, the decree bound them to assent interiorly that the Copernican theory was probably contrary to Holy Scripture. There was no question of absolute interior assent since the decree came from a fallible authority with no guarantee of inerrancy. This meant that if there were any Catholic astronomers who had no doubt that the Copernican system was true, they were excused from this internal assent. See Journet, *op. cit.*, p. 357. It is unfortunate that there still exists a need to repeat these fundamental doctrinal facts.

Kepler first heard about the prohibition. On August 4, 1619, Kepler wrote to Remus:

> The first I heard of my book [the *Epitome of Copernican Astronomy*] being prohibited in Rome and at Florence, was from your letter. . . . I pray you to send me the formula of censure . . . it means much to me to know whether the same censure will apply to Austria.[51]

It is well to remember that the Copernican system was still permissible as a hypothesis. Also, the opinion of the Consultors and the records of the proceedings remained hidden in the files of the Holy Office. The word "heretical" did not appear in the decree of the Index. Remus summed up the practical effect of the prohibition in a letter to Kepler dated August 13, 1619:

> I do not think that that book [Kepler's *Epitome*] will be prohibited except inasmuch as it may speak contrary to a decree of the Holy Office [actually, of course, it was a decree of the Congregation of the Index] of two years ago, or more. It was occasioned by a Neapolitan religious [Foscarini] who was spreading these opinions among the people by writing in the vernacular, from which were arising dangerous consequences and opinions, while Galileo, at the same time was pleading his cause at Rome with too much insistence. And thus Copernicus has been corrected, for some lines at least, in the beginning of his first book. But it may be read with permission, and so may the *Epitome* both by the learned and those versed in science, both in Rome and throughout Italy. There are no grounds for your uneasiness, either as regards Italia or Austria; only keep yourself within bounds, and put a guard on your feelings.[52]

[51] Cited by M. W. Burke-Gaffney, *Kepler and the Jesuits* (Milwaukee: Bruce, 1944), pp. 101 f.

[52] *Ibid.*, p. 102.

CHAPTER V

Galileo and Urban VIII

With the prohibition of Copernicus's work, the crisis seemed to be over. Theologians turned once again to writing refutations of Protestant doctrines and the Aristotelians went back to their texts. In July, 1616, Galileo left Rome and returned to Florence. He was discouraged and disappointed, but not defeated. There was always a chance that the decision might be reversed. Prohibitions had been repealed before. In 1590 a book entitled *Disputations on Controversies of the Christian Faith against the Heretics of this Age*, authored by none other than Robert Cardinal Bellarmine himself, had been placed on the Index of Forbidden Books by Pope Sixtus V. Pope Sixtus was enamored with the mediaeval idea of temporal rule by the Church and he resented Bellarmine's contention that the pope had no strictly temporal jurisdiction outside of the Papal States, but merely an indirect jurisdiction which could be exercized only when the actions of rulers threatened the eternal salvation of their subjects. Pope Sixtus ordered the *Controversies* to be prohibited, but died before the new edition of the Index was completely published in August, 1590. His successor, Pope Urban VII, who reigned for less than two weeks, immediately halted the presses and had the prohibition removed. Galileo had reason to hope that when Pope Paul V died, a successor would be elected who would erase the name of Copernicus from the Index.

Back in Florence, Galileo knew that he had to play it safe for the time being. There was no point in aggravating sensitive feelings in Rome. Besides, his health was poor and he needed a rest from the philosophical wars. Galileo busied himself by working on the grounds of his small estate, jotting down ideas and keeping in touch with his friends. It was a full two years before his pen spoke out again and then it was only with caution. In 1618, Archduke Leopold of Austria asked Galileo to send some of his writings to him. He was an amateur scientist and thought it would be interesting to correspond with someone of Galileo's reputation. Galileo saw the request as an opening, and, hoping that the Archduke would sponsor its publication, he forwarded a copy of his theory of the tides, the same theory he had asked Cardinal Orsini to present to the Pope two years before with drastic consequences. In a note accompanying the manuscript, he wrote ironic words of self-protection: "I consider this treatise which I send to you as merely a poetical conceit, or a dream ... because it is based on the double motion of the earth." [1] Leopold may have appreciated the manuscript, but he did not have it published.

The year 1618 brought with it the three great comets which heralded the beginning of the Thirty Years War. These comets were unusually bright and clearly visible for several months. Astrologers saw them as a sign that the world was about to end. And to many it seemed that this surely was the message which they were meant to convey to mankind. The religious and political unrest in Europe erupted into a full scale war in which the fighting was as bitter and bloody as any war has ever been. Many of the soldiers went into battle willing to kill or be killed in what they thought was a defense of their faith. Frequently, however,

[1] *Opere*, XII, 391.

it was political scheming rather than religious fervor which moved monarchs to commit their armies to battle. Thus Cardinal Richelieu of Catholic France joined forces with Gustavus Adolphus of Protestant Sweden and fought against the Holy Roman Emperor. There were those who welcomed the signs in the sky and who would not have been sorry if they were the signal that the chaos, killing, and the world itself were soon to end forever.

But astronomers took a different view of the comets. For one thing, it was the comets that revived interest in astronomy and celestial theories. Comets had always been a source of some consternation to astronomers and philosophers. Aristotle, because of his doctrine that the heavens are immutable, did not believe that comets were celestial phenomena at all. They were nothing more than earthly vapors rising in the atmosphere below the lunar sphere and ignited by the friction caused in the upper regions by the motion of this sphere. But Tycho Brahe had shown by painstaking observation of the great comet of 1577 that comets are really in the heavens, that is, they are beyond the supposed sphere of the moon which was the Aristotelian dividing line between the terrestrial and celestial regions. Comets were still mysteries, however. Their orbits did not seem to fit the paths of other celestial bodies, nor was their velocity constant. And no one knew where they came from.

A Jesuit Father, Horatio Grassi, delivered a series of lectures at the Roman College in which he explained that comets are real material bodies which undoubtedly move in the celestial region beyond the lunar sphere. These lectures were published early in 1619 under the title *Astronomical Discussion concerning the Three Comets of 1618*. When Galileo read this work, he felt he had to reply. Now Galileo could not account for the errant paths followed by comets and so held, incorrectly, that they were not in the heavens at all. He believed that they were merely optical phenomena produced by the refraction of the sun's ray in the upper atmosphere.

Galileo was in no position to offer a personal refutation of

Grassi. But there was a way to stir up a dispute. He accepted the offer of a friend and disciple, Mario Guiducci, to write up his arguments and publish them under Guiducci's name. Adopting this procedure, Galileo wrote two lectures which were delivered by Guiducci to the Florentine Academy and then published in June, 1619, under the title *A Discourse on Comets*. The battle was on and Galileo was once again in the thick of it. It was no secret that Galileo was the real force behind Guiducci and his friends rejoiced in public that he was making a comeback. Maffeo Cardinal Barberini, who had supported Galileo in the troubled days of 1616, wrote a poem in his honor entitled "Dangerous Adulation." But Barberini's secretary, Monsignor Ciampoli, wrote Galileo that while he was happy to see that the Florentine scientist would not accept a forced retirement, he thought it necessary to warn Galileo that the Jesuits were not at all pleased by his attack on one of their Company.

Grassi himself, copying Galileo's use of a pseudonym, came back with a refutation of the *Discourse on Comets*. It was titled *The Astronomical and Philosphical Balance* and published under the name of Lothario Sarsi. There is no doubt that Grassi baited Galileo in this pamphlet. He knew his opponent's handicap and he dared Galileo to come out in favor of the Copernican system. Galileo had suggested that the vertical path of the comet which veered off in a northerly direction might have to be explained by a further cause. Grassi jumped on this intimation:

> What is this sudden fear in an open and not timid spirit which prevents him from uttering the word which he has in mind? I cannot guess it. Is this other motion which could explain everything and which he does not dare to discuss—is it of the comet or of something else? It cannot be the motion of the circles, since for Galileo there are no Ptolemaic circles. I fancy I hear a small voice whispering discreetly in my ear: the motion of the earth. Get thee behind me thou evil word, offensive to truth and to pious ears! It was surely prudence to speak it with baited breath. For, if it were really thus, there would be nothing left of an opinion which can rest on no other ground except this false one.... But then certainly Galileo had no such idea, for I have never known him otherwise than pious and religious.[2]

[2] L. Sarsi, *Libra astronomica ac philosophica, Opere*, VI, 145 f.

One can hardly blame Galileo for accepting the challenge. He responded by writing what one scholar has called "the greatest polemic ever written in the history of science." [3] He began this work, the *Il Saggiatore*, in 1621. In January of that year, Pope Paul V died and was succeeded by Pope Gregory XV. Then Galileo's ruler, protector, and friend, Cosimo II, died, and because his son Ferdinand was too young to assume the throne, the rule of the Tuscan Republic fell to the regency of the Duchess Maria Madelaine. On September 17, 1621, the great and saintly Robert Cardinal Bellarmine passed away. Pope Paul, Cosimo, and Cardinal Bellarmine had all been important figures in the prohibition of 1616. Their deaths must have been noted with mixed emotions by Galileo.

Galileo spent two years working on *Il Saggiatore* which translated means *The Assayer*, thus implying that it would weigh things which were too fine for Sarsi's *Astronomical Balance*. Much of the work is taken up with Galileo's philosophy of science. He holds that natural philosophy may ask questions about the ultimate causes of reality, but it can never answer them with certainty. Only mathematics can provide the human mind with the certitude it seeks. What cannot be measured belongs to the province of faith or belief, but not, strictly speaking, to human knowledge. He insists that:

> Philosophy is written in this grand book, the universe, which stands continually open to our gaze. But the book cannot be understood unless one first learns to comprehend the language and read the letters in which it is composed. It is written in the language of mathematics, and its characters are triangles, circles, and other geometric figures without which it is humanly impossible to understand a single word of it; without these one wanders about in a dark labyrinth.[3A]

Galileo's true originality lay in "his insistence that the book of nature is written *only* in mathematical language." [4] For Galileo

[3] Stillman Drake, *Discoveries and Opinions of Galileo*, p. 227.

[3A] Galileo, *The Assayer*, trans. Drake, *op. cit.*, pp. 237 f.

[4] J. A. Weisheipl, O.P., *The Development of Physical Theory in the Middle Ages* (New York: Sheed and Ward, 1959), p. 84.

only a mathematical description of nature could really demonstrate nature's activity with a true, proper, and necessary explanation. Qualitative natural science had nothing to offer. Efficient and final causality, because they had no place in mathematical investigation, were regarded as idle speculation. In fact, "whatever could not be caught in mathematical abstraction, such as secondary sense qualities, essences, and causes, were either subjective or did not exist for Galileo." [5] Thus he could say:

> Hence I think that tastes, odors, colors, and so on are no more than mere names so far as the object we place them in is concerned, and that they reside only in the consciousness. [6]

What this amounts to is the divorce of science from philosophy. Natural philosophy with its explanations of nature based on concepts of act and potency, final causality, and nature as a principle of operation, is practically useless. True science, and for Galileo this means demonstrative knowledge, is mathematical only.

But *The Assayer* was not merely a treatise on science or an explanation of comets, it was also a sarcastic lampoon aimed at Father Grassi:

> Let Sarsi see from this how superficial his philosophizing is except in appearance. But let him not think he can reply with additional limitations, distinctions, logical technicalities, philosophical jargon, and other idle words, for I assure him that in sustaining one error, he will commit a hundred others that are more serious, and produce always greater follies in his camp . . . [7]

The frequency and sharpness of such remarks played a crucial part in alienating many Jesuits, whose cooperation Galileo so badly needed. Father Grienberger was to say in 1634, that, "Galileo should have known how to keep the affections of the fathers of the Roman College. If he had, he would still be living gloriously in the world, he would not have fallen into trouble, he would be able to write on any subject he wished, even the rotation of the earth." [8] This is probably somewhat exaggerated, but it is certain that the alienation of the Jesuits was to prove costly.

[5] *Ibid.*
[6] Galileo, *The Assayer, ed. cit.*, p. 274.
[7] *Ibid.*, p. 269.

In 1623, shortly before *The Assayer* was to be published, Pope Gregory XV died after a short reign of only two years. The man elected to succeed him was none other than Maffeo Cardinal Barberini, Galileo's old friend and defender.[9] This was the signal for Galileo's hope to turn into determination. Cardinal Barberini, now Pope Urban VIII, was a highly gifted man, well-versed in the arts, which, now that he was Pope, he could foster on a grand scale. Barberini had received his education from the Jesuits and, after his ordination to the priesthood, had risen quickly on the ladder of ecclesiastical success. In 1604 he was appointed Papal Nuncio to France and had handled that difficult assignment with such finesse that he received the cardinalate from Pope Paul V in 1606. Now, at the age of fifty-three, he was elected Supreme Pontiff. Pastor describes Pope Urban as a "self possessed and keenly obervant man who brooked no contradiction . . . he was slow to make up his mind and easily roused though also quickly calmed down." [10] A French Ambassador in Rome wrote of him in 1624:

> The Pope remains as he always was, sincere and frank, a friend of books and scholars, quick, fiery, somewhat choleric, impatient of opposition but ready to yield to solid arguments, full of the best intentions for the Church and Christendom.[11]

As a cardinal, Barberini had been very well disposed toward Galileo. He had written to the scientist during one of his illnesses:

> I write because men like you, who are of great value, deserve to live a long time for the public benefit, and I am also motivated by the particular interest and affection which I have for you, and by my constant approbation of you and your work.[12]

In a note accompanying the poem which he wrote in Galileo's honor, in 1620, Barberini wrote:

[8] *Opere*, XVI, 117.
[9] It is interesting to note that two other friends of Galileo were in the running before the Conclave elected Cardinal Barberini. Cardinals del Monte and Gaetani each received votes on several ballotings and both were instrumental in the eventual election of Barberini. See L. Von Pastor, *History of the Popes* (trans. E. Graf, London: Kegan Paul, 1938), XXVIII, pp. 1–23.
[10] *Ibid.*, p. 35.
[11] *Ibid.*, p. 37.
[12] *Opere*, XI, 216.

The esteem which I have always had for you because of your honor and virtues, has provided the matter for the enclosed composition.[13]

Galileo responded to the election of Pope Urban VIII by dedicating *The Assayer* to him. When this work was published, in October, 1623, it won the applause of Urban and the staunch cadre of Galileists perhaps more for the spirit reflected in it than for the doctrine which it proposed. But the diehard conservatives, both in theology and philosophy, some out of sincerity, others out of fear or envy, were still fighting for the old order. The fact that the Pope was an old friend of Galileo, that he had allowed *The Assayer* (which, after all, was an attack on a Jesuit) to be dedicated to him and had perused it without objection, caused a great deal of concern in the conservative camp. They must have wondered how long it would be before Galileo would somehow convince the Pope to revoke the censure of 1616 and support the new astronomy. They were well aware that if the decree of 1616 was ever erased, they would no longer be able to keep Galileo in check; they would not be able to ward off the new science.

Galileo himself thought the tide had turned. There was good reason to think so. A friend wrote from Rome saying that when he had visited the new Pope, nothing had given Urban so much joy as when the conversation turned to Galileo.[14] Prince Cesi reported that the Pope had asked him when Galileo would come to Rome. "...in a word, he showed that he loves and esteems you more than ever." [15] Galileo had heard enough. It was time to act and he knew it. Surely the Pope would be more willing to listen to an old friend than to the whisperings of his enemies. Perhaps Galileo could convince the Pope that by revoking the prohibition against Copernican writings, he could win himself an honored place in the history of science. In the spring of 1624 Galileo headed once more toward Rome.

[13] *Opere* XIII, p. 38.
[14] Renuccini to Galileo, October 20, 1623, in *Opere*, XIII, 139.
[15] Cesi to Galileo, October 31, 1623, *Ibid.*, 140.

Galileo's reception in Rome was heartening. Powerful cardinals such as Francesco Barberini, Cobelluzio, Boncompagni, and Frederick of Hohenzollern conversed with him at length on questions of astronomy and politics. The Pope presented him with two medals of gold and silver, a valuable painting, and a papal Brief addressed to the rulers of the Tuscan Republic recommending Galileo for further patronage because he was a man "... whose fame shines in the sky and is spread over the whole world." [16] On June 8, 1624, Galileo could report to Prince Cesi that Cardinal Hohenzollern had spoken to Urban about the Copernican question and had told him that German Protestants seemed very much in favor of the new system so that great caution should be exercized in any official decision on this matter. The Pope replied that the Church had not defined the Copernican opinion to be heretical nor would she ever do so.[17] In the same letter to Cesi, Galileo mentioned meeting a Dominican, Nicholas Riccardi, who was strongly of the opinion that the Copernican system had nothing to do with faith and that Holy Scripture had no place in this scientific dispute.[18]

Galileo himself had six long audiences with the Pope. It is unfortunate that we have no record of what was said during these talks. But we can reconstruct, with some probability, the tone and tack of these discussions. No doubt the Pope encouraged Galileo to continue his study and writing. Galileo probably complained that this would be difficult because his enemies were still talking of Copernicanism as though it were a heresy. Only an abrogation of the decree of the Index would be able to silence them and allow freedom of expression. Urban, while privately sympathetic to Galileo's request, had now to think of the whole Church. He must have remembered that although he had opposed the ban eight years before, there seemed to have been some good reason at least for issuing it. Furthermore, Roman authority was being ignored

[16] *Opere*, XIII, 183.

[17] Galileo to Cesi, June 8, 1624, *Opere*, XIII, 182.

[18] *Ibid.* This is important because Riccardi's sympathies for Galileo's plight will be partly responsible for the subsequent happenings.

and scorned enough already. A papal act revoking the Congregation's decree when no new evidence had been presented to force a reconsideration, would further diminish the authority of the Sacred Congregations in the public eye. However much Urban might have wanted to lift the prohibition and allow Copernicus to be openly defended, he felt it necessary to refuse on the grounds that authority was more necessary just then than was the new and unproven astronomy. Yet, recalling the tone of the prohibition, Urban conceded that so long as Galileo treated the Copernican theory as a hypothesis, he could write all he wanted on the subject. Just so he did not try to propose it as the only possible explanation of celestial motion. Urban, it seems, did not put much faith in human knowledge anyway. He was a relativist of sorts. As far as he was concerned, a strict demonstration of any system of the universe was impossible. This was not because he doubted that a parallax could ever be determined. His reasoning was that God, who is all powerful, could have established the machinery of the universe in such a way that no man could ever penetrate its mysteries. Thus even if all the evidence seemed to prove one system in preference to another, it could not be said to be demonstrated because true demonstration demands full certitude that the conclusion is true and cannot be otherwise. As far as Urban was concerned, this kind of certitude could exist only in the mind of God.[19] To put it simply, the Pope felt that natural philosophy and applied Physics could conclude only in belief, never in certitude. It is true that as a lover of the arts and sciences, Urban wanted Galileo to continue his work. But by refusing to revoke the censure of 1616 he made it quite clear that Galileo would have to exercize prudence and confine himself to the realm of hypothesis. It was good that Galileo wanted to write, but while a dialectic would be all right, an attempt at a demonstration would not. Because of this ambiguous notion of scientific knowledge and hypothesis, Urban failed to clarify for Galileo just what he could or could not write.

It is not hard to imagine Galileo blinking in disbelief at the

---

[19] See Galileo's letter to Cesi already cited in note 17.

Pope's relativism. But he probably thought to himself that he could use this strange epistemology to his own advantage, for in ambiguity at least there is leeway. It seemed to Galileo that if there was a practical conclusion to his talks with the Pope it was that he could write whatever he pleased so long as he admitted that his theory might not be the final answer. This meant that he could even commit himself fully to the Copernican system in the body of his text and then submit to the Pope's wishes by claiming in the preface and conclusion that his doctrine might not be true because, after all, who could know the mind of God? In his way he could have his cake and eat it too. He could aim his whole line of argument at producing a demonstration and at the same time fend off the theologians with cleverly worded submission clauses. Galileo returned to Florence in June, 1624, confident that even though he had not been able to get a revocation of the prohibition, he had enough freedom without it.

He tested the idea of a formal submission clause by writing an open letter to Francesco Ingoli, who eight years before had published a pamphlet containing all the objections against the Copernican system. Galileo's letter was an open defense of Copernicanism in scarcely-veiled terms of established fact. In a somewhat esoteric fashion he claimed that:

> I further contend that I know other facts from experience which as yet have been observed by no one else; from these facts, within the limitations of human considerations, the truth of the Copernican system seems certain.[20]

But in the first part of the letter, he had protected himself by saying that he presented the arguments for the Copernican theory in order to show the Protestants that it was not out of ignorance that the Church had banned the heliocentric doctrine, but out of reverence for Holy Scripture and the Fathers, which carry more authority than does human reason. The *Letter to Ingoli* was never published. It did not need to be. Copies were made and circulated freely. The Pope had it read aloud and expressed enthusi-

[20] *Opere* VI, 543.

asm.[21] Several high officials studied it and no complaint was forthcoming. The submission gimmick had worked. Galileo's hopes were further raised early in 1625 when he received word that his *Il Saggiatore* had been denounced to the Holy Office but that Father Guevara, the Consultor assigned to examine the work, praised it in his report and said that even the doctrine of the earth's motion did not seem to him to deserve condemnation. The Holy Office accepted his opinion and the matter ended there.

Encouraged by these events, Galileo went back to work on the book which would present what he considered the final and certain proof that Copernicanism was a fact, his monumental *Dialogue on the Great World Systems*. While most of the arguments which he was incorporating into the *Dialogue* were not new, he intended to present them in an extended and final form. Perhaps if the *Dialogue* had followed closely on the *Letter to Ingoli*, subsequent events would have taken a different course. But frequent illness and other circumstances hindered his progress on the book. From 1626–1629 he was able to do almost no work on it. The delay was unfortunate for it allowed the opposition of philosophers and theologians to ferment. They knew that Galileo was pushing his doctrine again and that this time he had a sympathetic listener in the person of the Pope. Furthermore, Father Nicholas Riccardi, who was favorable to Galileo, had just been appointed to the influential post of Master of the Sacred Palace.[22] Things began to look too much in Galileo's favor to suit his enemies. And their fears that Galileo was about to start another pressure campaign for the acceptance of Copernican astronomy were not without foundation. There can be no doubt of what Galileo's intention was in writing the *Dialogue*. He confided to a friend, Elia Diodati, on October 29, 1629, "I have taken up work again on the *Dialogue . . . it will provide I trust, a most ample confirmation of the Copernican system.*" [23]

Perhaps if the arguments in the *Dialogue* had been strong

[21] *Opere* XIII, 295.
[22] As we have seen, Galileo met Riccardi in 1624 and knew that Riccardi's sentiments favored his right to freedom of expression.
[23] *Opere*, XIV, 49.

enough he actually would have been able to win Church approval. Armed with better proof, Galileo's powerful friends in Rome might well have been able to convince the Pope that, relative or not, Copernicanism was the best formulation of celestial motions yet devised by man and that the Church should commit herself to this doctrine at least by allowing the alternative exegesis of the pertinent scriptural passages.

But the *Dialogue*, while devastating in its criticism of Aristotelian physics and Ptolemaic astronomy, did not have much direct proof to offer in favor of the new astronomy.[24] It was constructed around a Socratic-type conversation between three main characters. Salviati speaks for Galileo. Sagredo is the supposedly impartial listener. Simplicio is the simple Aristotelian whose mind is cobwebbed by the cosmology of his Master. Four "days" of discussion divide the *Dialogue*.

On the first day, the reader is urged, by the force of the discussion, to abandon once and for all the distinction between celestial and terrestrial matter, a distinction basic to the whole Aristotelian cosmology. Sunspots, variable speeds of heavenly bodies, and mountains on the moon all combine to show that celestial bodies are not the perfect, uniform unchanging spheres of Aristotle's cosmos; instead they are of the same mutable nature as the earth. Salviati admits that this mistaken doctrine is not altogether Aristotle's fault; after all, he did not have a telescope. But with contemporary Aristotelians there could be no question of invincible ignorance and Salviati points out to Aristotle's followers how much they lack the spirit of their Master:

---

[24] It is interesting to compare the views of Galileo's *Dialogue* as presented in two recent studies on the Galileo case. Giorgio de Santillana (*op. cit.*, p. 175) writes, "A modern may miss the intellectual tension of abstract developments, the tightness of the formula which brings the theory into shape. But Galileo is willing to pay that price in order to remain a man among men, a person and a force within his own culture." Arthur Koestler (*The Sleepwalkers*, p. 478) says: "There can be no doubt that Galileo's theory of the tides was based on unconscious self-deception; but . . . there can also be little doubt that the sunspot argument was a deliberate attempt to confuse and mislead . . . We have seen that scholars have always been prone to manias and obsessions, and inclined to cheat about details; but impostures like Galileo's are rare in the annals of science."

> I affirm that we have in our age new occurences and observa-
> tions and such that I doubt not in the least that, if Aristotle
> were here today, they would make him change his opinion. This
> may be easily gathered from the very way he argues, for when
> he writes that he esteems the heavens unalterable because no
> new thing was seen to be born there, or any old one to be
> dissolved, he seems to imply that, if he were to see any such
> accident, he would then hold the contrary and put observation
> before natural reason (as indeed is right) for, had he not made
> any reckoning of the senses, he would not then have argued
> immutability from not seeing any change.[25]

And Sagredo agrees by attacking the doctrine that heavenly
bodies are more perfect because they do not change:

> I cannot without great wonder, nay more, disbelief, hear it being
> attributed to natural bodies as a great honor and perfection that
> they are impassible, immutable, inalterable, etc.: as, conversely,
> I hear it esteemed a great imperfection to be alterable, generable
> and mutable. It is my opinion that the earth is very noble and
> admirable by reason of the many and different alterations, mu-
> tations, and generations which incessantly occur in it. And if,
> without being subject to any alteration, it had been all one vast
> heap of sand, or a mass of jade, or if, since the time of the
> deluge, the waters freezing which covered it, it had continued
> an immense globe of crystal, wherein nothing had ever grown,
> altered or changed, I should have esteemed it a wretched lump
> of no benefit to the universe, a mass of idleness, and in a word
> superfluous, exactly as if it had never been in nature.[26]

Now to unify celestial and terrestrial physics, Galileo has to show
that the circular motion observed in the heavens is not essentially
different from the motion of objects on earth, and by doing this
he will also remove an objection against the possibility of the
motion of the earth itself. He does this not by denying the natural
circular motion of planets, for that would have meant postulating
an infinite universe.[27] Instead, he makes inertial motion circular

[25] Galileo, *Dialogue on the Great World Systems*, trans. G. de Santillana
(Chicago: Chicago University Press, 1953), p. 59.

[26] *Ibid.*, p. 68.

[27] Professor A. Koyré maintains that Galileo never really decided whether
the universe was finite or infinite: "Some features of his dynamics, the fact
that he could never free himself from the obsession of circularity . . . seem to
suggest that his world was not infinite. If it was not finite, it was probably,
like the world of Nicholas of Cusa, indeterminate." Koyré, *From the Closed
World to the Infinite Universe* (New York: Harper Torchbooks, 1958), p. 99.

and, contrary to his own experimental knowledge, declares that free falling bodies describe a circular path. Speaking through Salviati, Galileo affirms that:

> ... whatever moves with a straight motion changes place and, continuing to move, moves by degrees farther away from the term from whence it departed and from all the places through which it has successfully passed. If such motion naturally suited with it, then it was not in the beginning, in its proper place; and so the parts of the world were not disposed with perfect order. But we suppose them to be perfectly ordered; therefore, as such, it is impossible that by nature they should change place and consequently move in a straight motion. Moreover, the straight motion being by nature infinite, because the straight line is infinite and indeterminate, it is impossible that any movable body can have a natural principle of moving in a straight line, namely, toward the place whither it is impossible to arrive, there being no predetermined limit; and Nature, as Aristotle himself well says, never attempts to do that which cannot be done or to move whither it is impossible to arrive.[28]

What this means is that Galileo's attempt to unify the physics of the heavens and earth is successful in that it postulates some kind of inertial motion. But it is completely wrong in making that motion circular rather than in a straight line. Unable to extricate himself from the dogma of the circle which for centuries misled astronomers, he cannot come forth with a concept of an infinite universe, and has to make do with what he has.

> I answer that none of the conditions whereby Aristotle distinguishes the celestial bodies from the elementary has any foundation other than what he deduces from the diversity of their natural motions; so that, if it is denied that the circular motion is peculiar to celestial bodies, and affirmed instead that it is agreeable to all naturally movable bodies, one is led by necessary consequence to say either that the attributes of generated or ungenerated, alterable or unalterable, partable or unpartable, etc., equally and commonly apply to all bodies, as well to the celestial as to the elementary, or that Aristotle had badly and erroneously deduced those from the circular motion which he has assigned to celestial bodies.[29]

[28] *Ibid.*, pp. 23 f.
[29] *Ibid.*, p. 45.

A stone dropped from a tower thus undergoes two motions simultaneously. It falls downward, and it shares in the west-east motion of the earth. To an observer outside the earth these two motions would combine to describe a circular path. It is true that Galileo will alter this position in a later work, the *Discussion on Two New Sciences*, but as Professor Dijksterhuis points out:

> ... there are actually several Galileos, and the difficulty of reading the *Dialogue* is due in particular to the fact that in this work they sometimes speak together. But one thing should have become clear: in the passages hitherto discussed there is no question of the conception of inertia that is formulated in Newton's first law.[30]

The remainder of the first day is spent in a discussion of many basic Aristotelian concepts, the idea being that Galileo must undermine the Aristotelian system, must challenge its fundamental premises if he is to convince open-minded readers that a change is necessary.

The second day focuses on the possibility of a moving earth and the reasons which seem to favor such an astronomy as not only possible but true. The discussion begins with a lengthy attack against the text-citing Aristotelians. Salviati speaks for Galileo:

> I have many times wondered how these pedantic maintainers of whatever came from Aristotle's pen are not aware how great a prejudice they are to his reputation and credit and how, the more they go about to increase his authority, the more they diminish it. When I see them obstinate in their attempts to maintain those propositions which are manifestly false, and trying to persuade me that to do so is the part of a philosopher, and that Aristotle himself would do the same, it much discourages me in the belief that he has rightly philosophized about other conclusions, for if I could see them concede and change opinion in a manifest truth, I would be more willing to believe that, where they persist, they may have some solid demonstrations, by me not understood or even heard of.[31]

[30] E. J. Dijksterhuis, *The Mechanization of the World Picture* (Oxford: Clarendon Press, 1961), p. 351.
[31] Galileo, *The Dialogue on the Great World Systems*, ed. cit., p. 124.

The discussion now moves to the issue at hand. Salviati proposes one of the main arguments in favor of a moving earth:

> If we consider only the immense magnitude of the starry sphere compared to the smallness of the terrestrial globe, and weigh the velocity of the motions which must, in a day and night make an entire revolution, I cannot persuade myself that there is any man who believes it more reasonable and credible that it is the celestial sphere that moves around, while the terrestrial globe stands still.[32]

After hearing a number of variations on this theme: that it seems unnecessary to make the entire universe race around the earth when one turn of the earth could accomplish the same effect, Simplicio answers that all of Salviati's arguments are tending toward the same term:

> It seems to me that you base all you say upon the greater simplicity and facility of producing the same effects. To this I reply that I am also of the same opinion if I think in terms of my own not only finite but feeble power; but, with respect to the strength of the Mover [God], which is infinite, it is no less easy to move the universe than the earth, yea, than a straw.[33]

Salviati sidesteps this answer neatly by quoting the Aristotelian maxim, *frustra fit per plura quod fieri potest per pauciora*—"That is done in vain by many which can be done by fewer." Galileo is making an issue of simplicity. Yet, in reality, the Copernican system was only a slight gain in simplicity over the Ptolemaic. Both were encumbered with epicycles. Only Kepler had been able to free astronomy of epicycles, but Galileo, though he knew of Kepler's laws never bothered to adopt them.

The second day continues with a consideration of physical objections against a moving earth. Simplicio points to the fact that a heavy body falling downward moves in a line perpendicular to the surface of the earth. Now if a stone is dropped from the top of a tower and the earth is moving, the stone will land some distance to the west of the tower rather than at its base because

[32] *Ibid.*, p. 128.
[33] *Ibid.*, p. 136.

the tower is attached to the moving earth while the falling stone is not. The same effect would be achieved, so the objection goes, by dropping a stone from the mast of a ship. If the ship is not in motion, the stone will fall to the foot of the mast. But if the ship is in motion, the stone will land away from the base of the mast. Galileo denies the premises and the conclusion. He correctly states that the stone will fall to the base of the mast whether the ship is moving or not for the simple reason that if the ship is moving, the stone shares the ship's motion. Similarly bodies detached from the earth would retain the velocity received from the earth's rotation and thus "one cannot, from the stone's falling perpendicularly at the foot of the tower, conclude anything touching the motion or rest of the earth." [34]

Another argument against the mobility of the earth stated that men, animals, and buildings would be flung off into space by a whirling world. Concerning this difficulty and Galileo's handling of it, Professor Dijksterhuis remarks:

> This is, of course, a very sound argument because, as we now know, there does exist a velocity of rotation at which the force of gravitation would not be able to provide the centripetal acceleration that terrestrial bodies require for their axial rotation. Nor can this argument be refuted without any quantitative formulation, merely by talking about it. However, that is precisely what Galileo tried to do; he asserted that things were not in danger of being flung away, and that consequently in this respect, too, everything remained exactly the same on a rotating earth as it would on a fixed earth.
>
> This is all the more curious because he was quite well aware that a body swung round on a string exerts an outwardly directed force on the string, and because he was no doubt able to observe that at a certain velocity of revolution the body tears itself away or the string breaks. However it seems that the contradiction between this fact and his statement about the absence of the same effect in the case of the rotation of the earth did not occur to him at all.[35]

The discussion moves on to such topics as projectile motion, the flight of birds, gravity, and the role of the senses in physical

[34] *Ibid.*, p. 159.
[35] E. J. Dijksterhuis, *op. cit.*, p. 357.

knowledge. As the day draws to a close Simplicio is moved to say:

> ... the discourses that this day have come under our debate have appeared to me fraught with very acute and ingenious notions, alleged by Copernicus and his side, in confirmation of the motion of the earth; but yet I do not find myself persuaded to believe it. For, in short, the things that have been said conclude no more than this: that the reasons for the stability of the earth are not necessary; but all the while no demonstration has been produced on the other side that does necessarily convince and conclude for its mobility.[36]

On the third day of the *Dialogue,* the participants weigh the evidence relative to the two main systems of the universe. Galileo does not think that the Tychonic alternative is even worthy of consideration. That, as J. L. E. Dreyer has noted, was the biggest single mistake of the *Dialogue.* Galileo must have been aware that many influential thinkers, including the Jesuit astronomers, had retreated to the Tychonic system as a halfway house until evidence could merit a complete break with any form of geocentric universe. Had he dedicated a day of the *Dialogue* to showing the inadequacies of the Tychonic system, he undoubtedly would have been in a better position to argue for a heliocentric astronomy. Instead, Galileo had to resort to the same old appeal: since the Copernican system saves the appearances better than the Ptolemaic, it must be true. His argument is that the planets are revolving around the sun and not the earth as their center. This implies that if the earth is not the center of the universe, it must be a mere planet revolving around the center, the sun. There is no mention of the fact that Tycho's system accounts for these planetary motions too. But when Sagredo asks why—if the Copernican system is so obviously superior—doesn't it have more adherents, he is told:

> Small regard, in my judgment, ought to be had of such thick souls as think it a most convincing proof to confirm and steadfastly settle them in the belief of the earth's immobility to see

[36] *Dialogue, ed. cit.,* p. 288.

that in the same day they cannot dine at Constantinople and sup in Japan, and that the earth, as being a most grave body, cannot clamber above the sun and then slide headlong down again. Of such as these, whose number is infinite, we need not make any reckoning, nor need we to record their fooleries or strive to gain to our side, as partakers in subtle and difficult opinions, individuals in whose definition the kind only is concerned and the difference wanting. [By definition, man is a rational animal. Take away the specific difference "rational" and you are left with Galileo's opinion of his opponents.] Moreover, what ground do you think you could be able to gain, with all the demonstrations of the world, upon brains so stupid as are not able of themselves to know their utter follies? ... You wonder that so few are followers of the Pythagorean opinion; and I am amazed how there could be any yet left till now that do embrace and follow it. Nor can I sufficiently admire the eminence of those men's intelligence who have received and held it to be true, and with the sprightliness of their judgments offered such violence to their own senses that they have been able to prefer that which their reason dictated to them to what sensible appearance represented most manifestly on the contrary.[37]

The discussion next moves to a consideration of sunspots, one of the two crucial "proofs" which Galileo tried to adduce in favor of the Copernican system. The argument from sunspots is not so simple as some have thought.[38] But neither is it conclusive. Salviati, Galileo's spokesman, observes that spots can be seen to pass over the face of the sun in such a way as to describe an arc which appears sometimes to turn upward, sometimes downward.

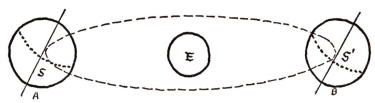

When the earth is at position A, the sunspots appear to incline upward. From position B, they seem to incline downward. Since this is true, Galileo argues, the earth must move around the sun.

---

[37] *Ibid.*, p. 341.
[38] For example, Arthur Koestler (*The Sleepwalkers*, p. 478) definitely underplays the cogency of the argument. See S. Drake and G. de Santillana, "Koestler and His Sleepwalkers," *Isis*, L, n. 3 (September 1959), p. 257.

This can be explained if it is granted that the axis of the sun is inclined to the ecliptic and the spots are seen from the earth as it moves from one side of the sun to the other.

Simplicio answers by saying that Salviati's argument is right but that the same rise and descent of the sunspots can be explained even if the earth is immobile and the sun moves around it. And he is correct.

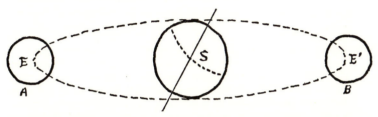

If the sun moves, when it is at position A, the spots will appear to a viewer on earth to slant downward, when at position B, to slant upward.

Salviati sees the strength of Simplicio's observation and replies that if the earth were stationary, it would be impossible to account for the continual variations in the tilt of the sun's axis necessary to cause the yearly variation in sunspot paths.[39] As Taylor comments, "This explanation is far from clear, and it is impossible to see why Galileo, whose opinion Salviati expresses, came to such a conclusion.[40]

Toward the end of the third day, Galileo admits that there still exists a solid objection to the heliocentric hypothesis: the lack of a stellar parallax. In what must have looked to his readers as another one of those "subtle and difficult opinions" Galileo is forced to give the same answer which Copernicus had given a century before:

> ... the immense distance of the starry sphere [possibly] renders such very small phenomena unobservable; these phenomena, as I

[39] See Stillman Drake, "A Kind Word for Sizi," *Isis*, II, n. 2, (June, 1958),
[40] F. S. Taylor, *Galileo and the Freedom of Thought*, p. 135. The illustrations for the arguments from sunspots are adapted from Taylor's book, p. 134.

have said, may possibly not have been hitherto so much as sought for, or, if sought for, yet not sought for in such a way as they ought, to wit, with that exactness which would be necessary to such a minute variation; and this exactness is very difficult to obtain, as well by reason of the deficiency of astronomical instruments, subject to many alterations, as also through the fault of those who handle them with less diligence than is requisite.[41]

It is on the fourth and final day that Galileo proudly unveils his "clinching proof," his theory of the tides. Here, Galileo felt, was a physical effect which could have only one necessary cause: the double motion of the earth.[42] He introduces it by having Simplicio declare:

> The introduction of the earth's motion and making it the cause of the ebbing and flowing of tides seem to me as yet a conjecture no less fabulous than the rest of those I have heard; and if there should not be proposed to me reasons more conformable to natural matters, I would without any more ado proceed to believe this to be a supernatural effect, and, therefore, miraculous and unsearchable to the understanding of men, as infinite others there are that immediately depend on the omnipotent hand of God.[43]

This sounds a great deal like Pope Urban's relativism and occultism. Of course the tides are not miracles and Galileo assures poor Simplicio that there is no need to "run to miracles" to explain them.

Galileo's theory of the tides is based on the analogy of a water barge bringing fresh water to Venice. As the barge starts or stops, the water contained in it piles up at one end, then the other. So it is with the seas, for the floors of the seas move with the earth. A. C. Crombie gives a clear summary of the complicated argument:

> Suppose the earth to be rotating on its axis and also moving in its orbit around the sun. One side of the spinning earth will now have a velocity equal to the sum of the velocities of rotation and

[41] *Dialogue*, ed. cit., p. 397.
[42] This was the same proof which Galileo had tried to use in 1616 without success and which he sent to Archduke Leopold of Austria in 1618.
[43] *Dialogue, ed. cit.*, pp. 429 f.

of movement around the orbit, while the other side will have a velocity equal to the difference of the two velocities. The earth is therefore moving with a greater absolute motion (relative to the fixed stars) on one side than on the other, and therefore the water of the seas should be left behind by the sea floor as the earth accelerates from the slow to the fast side, and should catch up on the sea floor as the earth slows down from the fast to the slow side. . . . It is easy now to see that the fallacy is that a uniform motion of progression (relative to the fixed stars) of the whole globe would not affect the relation of its parts to one another . . . [44]

Galileo rejected Kepler's explanation that the tides are caused by the moon's attraction.[45] Again a willingness to study Kepler might have saved him embarrassment. But he was captivated by the necessity of finding a physical proof and he convinced himself that he had it in the tidal theory. As far as his readers were concerned, Galileo's explanation of the tides was even less tenable than the Copernican system itself. The fact is that Galileo's tidal theory was far too imperfect and too subtle to merit the certitude with which Galileo proposed it as a panacea for Copernican problems.

Though he tried, Galileo did not prove that the earth moves. He did succeed in showing that it is not impossible for the earth to move; his arguments against the objections to such motion were, in general, quite valid. This in itself was a great accomplishment, enough in fact to merit the *Dialogue* an important position in the history of science. But it was not enough for Galileo who wanted to be remembered as the man who proved that the earth moves. Historians will always be able to ask what fate the *Dialogue* might have met had he really demonstrated his thesis. As it was, he had again committed himself to his theory as though it were fact and tried to convert his suspecting, conservative, but not

[44] A. C. Crombie, "Galileo's *Dialogue concerning the Two Principal Systems of The World*," *Dominican Studies*, III, (April–June, 1950), p. 132.

[45] ". . . in rejecting the moon's 'dominion over the waters' he flew in the face of an ancient tradition firmly rooted in observation. Galileo flouted tradition elsewhere and won; in his theory of the tides he lost what was perhaps his greatest gamble" H. L. Burstyn, "Galileo's Attempt to Prove that the Earth Moves," *Isis*, LIII, n. 2, (June, 1962) p. 181.

totally ignorant contemporaries. True, he did insert a submission clause at the end of the book. It was Urban's maxim that God could have established the universe in such a way that man could never discover its secrets and therefore to hold one system as certain would be to place limitations on the power and wisdom of God. But Galileo had the objection spoken by the poor Aristotelian, Simplicio, who had been buffeted about during the entire four-day discussion:

> As to the discourses we have held, and especially this last one concerning the reasons for the ebbing and flowing of the ocean, I am really not entirely convinced; but from such feeble ideas of the matter as I have formed, I admit that your thoughts seem to me more ingenious than many others I have heard. I do not therefore consider them true and conclusive; indeed, keeping always before my mind's eye a most solid doctrine that I once heard from a most eminent and learned person, and before which one must fall silent, I know that if asked whether God in His infinite power and wisdom could have conferred upon the watery element its observed reciprocating motion using some other means than moving its containing vessels, both of you would reply that He could have, and that He would have known how to do this in many ways which are unthinkable to our minds. From this I forthwith conclude that, this being so, it would be excessive boldness for anyone to limit and restrict the divine power and wisdom to some particular fancy of his own.[46]

It does not help matters any that Salviati answers:

> An admirable and angelic doctrine, and well in accord with another one, also divine, which, while it grants to us the right to argue about the constitution of the universe (perhaps in order that the working of the human mind shall not be curtailed or made lazy), adds that we cannot discover the work of His hands. Let us, then, exercize these activities permitted to us and ordained by God, that we may recognize and thereby so much the more admire His greatness, however much less fit we may find ourselves to penetrate the profound depths of His infinite wisdom.[47]

Coming as it did at the conclusion of a strong stand for the

[46] *Dialogue*, trans. Stillman Drake (Berkeley: U. of California Press, 1962), p. 464.
[47] *Ibid.*

Copernican system, this looked very much like an inane, futile, face-saving dodge of pietism which had to be used to escape the bonds of cold, hard logic. It stands in silent contradiction to the entire text.

The *Dialogue* was completed during January, 1630. Now the problem was to obtain official permission to publish it. Letters from Rome assured Galileo that all was well. Castelli reported on March 16, that the Pope had told Campanella (who was released from the fortress of Naples in 1626 at Urban's request) that "it [the prohibition] was never my intention; and if it had been up to me, that decree would never have been passed." In May, 1630, feeling that he had both the proof and the security he needed, Galileo went to Rome to see about publication.

Niccolo Riccardi of the Order of Preachers was the Master of the Sacred Palace and was thus entrusted with the task of approving or forbidding the publication of books in Rome. He was a delightful, portly friar who had been nicknamed "Padre Mostro," "Father Monster," by King Philip III of Spain.[49] A fellow-Florentine and an exceptionally capable theologian, Riccardi was sympathetic toward Galileo's cause. But he was also conscientious. However much he might have wanted to dispense with the technicalities, he had to live up to the responsibilities of his office. The problem facing him was not an easy one. He knew that Pope Urban himself had encouraged Galileo to write. Monsignor Ciampoli, Urban's secretary and Galileo's friend, probably reminded Riccardi every time he saw him that it was the desire of the Pope to see the *Dialogue* in print. But Riccardi also knew that the new astronomy was not to be presented as demonstrated and certain. The Pope had not revoked the decree of 1616 and that decree would have to serve as his norm in judging the book. Riccardi was not unaware, either, that there was a considerable army of men in Rome, many of them Jesuits who were angry at Galileo

---

[48] Castelli to Galileo, March 16, 1630, *Opere*, XIV, 88.
[49] See R. Mortier, *Histoire des Maitres Généraux de L'Ordre des Frères Prêcheurs* (Paris: Pichard, 1913), VI, p. 301.

for the way he had attacked Fathers Scheiner and Grassi of the Society, who were ready to pounce on the book if it was not altogether hypothetical. In short, it was a touchy affair.

Riccardi read the *Dialogue* and had his doubts about its mode of presentation. He passed it on to his assistant, Father Raphael Visconti, for his opinion. Visconti changed a few words and gave it back saying that he thought it was safe enough to be published. Riccardi still was not satisfied. Monsignor Ciampoli and Francesco Niccolini, who had replaced Guicciardini as the Tuscan Ambassador in Rome and whose wife was, by happy coincidence, Riccardi's cousin, began to exert pressure on the worried Master of the Sacred Palace. They convinced him that further delay was unnecessary and that he should grant the *Imprimatur* so that printing could begin immediately. Riccardi finally agreed to give the necessary license to the printer but with the stipulation that he would keep the manuscript. As soon as he had finished reading and correcting each page, he would send it to the printer. Meanwhile, he insisted, Galileo would have to rewrite the preface and conclusion and make them more in accord with the Pope's wishes. Galileo accepted these conditions, and Riccardi granted the *Imprimatur* for Rome. When Galileo left for Florence in late June it seemed that the fates were with him after all.

Prince Cesi, the Founder and President of the Lincean Acadamy, died shortly after Galileo left Rome. He had been a devoted friend and ally to Galileo through the years and his death must have saddened the scientist. But coming as it did at a critical time when Cesi was supposed to oversee the complex printing operation in Rome, the Prince's death was nothing short of tragedy. To make matters worse, a sudden outburst of the plague in Italy threatened to break off, or at least seriously hinder the communications between Florence and Rome. A letter arrived from Father Castelli urging Galileo for very serious reasons which he did not wish to record in writing, to ask permisson to have the *Dialogue* published in Florence as soon as possible. Galileo, not knowing what these serious reasons were and wondering if something had gone wrong in Rome, wrote to Niccolini and asked him to work on securing the necessary permissions from Riccardi.

The request was quite contrary to ordinary procedure and Father Riccardi at first refused to grant it. But once again the Ambassador managed to get his way. Riccardi agreed to turn the body of the text over to Giacinto Stefani, a Consultor of the Inquisition in Florence, but he, Riccardi, would keep the preface and conclusion himself. He then sent the text to Stefani along with a note of caution:

> I remember though that the mind of the Holy Father was that the title and subject ought not to propose the ebb and flow of the sea, but strictly the mathematical considerations of the Copernican position on the earth, with the intention of proving that, abstracting from God's Revelation and Sacred Doctrine, all the appearances could be saved by this position, undoing all the contrary arguments which can be brought forth by experience and Peripatetic philosophy, so that absolute truth is not conceded, but only hypothetical truth, and without Sacred Scripture, to this opinion. He must also show that this writing only proves that all the reasons which can be brought forth in favor of this position are known and that it is not from ignorance of the evidence that this opinion is banned in Rome ... with this caution, the book will have no difficulty in Rome.[50]

Stefani read the *Dialogue* and changed only a few words. Riccardi then asked Father Clement Egidii, the Head Inquisitor of Florence to make sure that the treatment did not go beyond the hypothetical. Egidii also approved the book. It is probable that both Stefani and Egidii did not read the work too closely or worry about granting an *Imprimatur* for its publication in Florence. After all, Riccardi had already granted it an *Imprimatur* in Rome.

Riccardi still had the revised preface and conclusion. But he sensed that no matter what changes were or were not made, the book was going to cause trouble. He delayed as long as possible. Niccolini and Ciampoli pressed him again. Finally, on July 19, 1631, Niccolini wrung the preface and conclusion from the hands

[50] *Opere*, XIX, 327.

of the fretting Riccardi and the *Dialogue* went to press. In February, 1632, printed copies were available to the public.

The *Dialogue* caused a storm. Theologians, philosophers, and laymen alike recognized the book as a well-camouflaged but real defense of the Copernican system. The contrast between the preface, which claimed that the arguments would be studied in an atmosphere of impartiality, and the text itself—which was tilted completely in favor of Copernicus—was patent even to amateurs. Riccardi was sorry he had ever yielded to pressure and allowed the publication of the book. To Galileo's friends, however, the *Dialogue* represented a masterful thrust at the old system. The irrepressible Friar Campanella congratulated Galileo:

> Each character plays his part wonderfully: and Simplicio seems the laughing stock of this philosophical comedy, for he shows the complete folly of his position: the verbiage, instability, and ostentation of it. We do not need to envy Plato. Salviati is a great Socrates who plays the midwife to others' brainchildren more than he bears his own, and Sagredo is a liberal master, who without having been bastardized by the schools, judges everything with great wisdom.[51]

But even Campanella had some reservations. Referring to Galileo's doctrine that the earth's motion is the cause of the tides, he wrote, "About the movement of the sea, I am not altogether with you, although it was better in the writen version than it sounded when I heard it from my friends." [52] But Campanella was wise enough to see the seeds of a new science in Galileo's book and he predicted, "These truths of ancient novelties, of new worlds, new stars, new systems etc. . . . are the beginning of a new age." [53]

Not many theologians were as open-minded as Campanella. Some undoubtedly feared that if this open breach of a Roman Congregation's decree went unpunished, the precedent would weaken the effect of all subsequent decrees. Others were angered because their doctrinal position, represented by Simplicio, was

[51] *Opere*, XIV, 366.
[52] *Ibid.*, 367.
[53] *Ibid.*

mocked and scorned. Riccardi saw the handwriting on the wall. "The Jesuits," he said, "will persecute him [Galileo] most bitterly." [54]

Everyone knew that any official action against Galileo would have to have the Pope's approval. Someone who knew Urban's weak points went to work on him. Urban was shown quotations from the *Dialogue* and told that they were intended to mock him and his ideas. Had not Galileo put the Pope's argument that God was all powerful and therefore no system of the universe devised by man could claim any certitude—in the mouth of Simplicio? Urban's vanity was greatly offended. He began to suspect that throughout the book Simplicio was meant to be a caricature of Urban. And when it was pointed out that the arguments presented in the *Dialogue* for the Copernican system could not, by any standard, be called impartial or hypothetical, the Pope was sure that he had been betrayed. Riccardi, called into private audience, told him the whole story of the *Imprimatur*. Urban was convinced that his kindness in encouraging Galileo to publish had been repaid with treachery. He decided that the fact that an *Imprimatur* had been given was primarily the fault of Monsignor Ciampoli, since it was he who had assured the Pope that the papal instructions had been carried out and had given Riccardi direct clearance to license the book. Ciampoli was relieved of his duties as papal secretary.

Galileo, for his part, had taken a calculated risk. He probably wouldn't have tried to publish if he had not thought that with the ascent of Urban VIII to the pontifical throne, the cultural atmosphere in Rome had grown more favorable. And, recalling Barberini's enthusiastic support in earlier years, he had every right to hope so. He gambled that by presenting the glaring defects of the Ptolemaic system and defending the logic of the Copernican theory under a guise of feigned neutrality, he could compel the Church to see its mistake, revoke the prohibition, and adopt the

[54] Riccardi, quoted by F. Magalotti in a letter to Mario Guiducci, August 7, 1632, *Opere*, XIV, 370.

new astronomy. Though perhaps not entirely from religious motives, Galileo wanted the Church to be the patron and protector
of the new science.

Galileo had no intention of ridiculing the Pope or of destroying
the doctrinal authority of the Roman Congregations. But the
style is the man and Galileo was by temperament a polemicist. It
was not Galileo's fault that some of the arguments used in support of the old system looked pretty bad on close inspection. However it is true that the Pope's argument was poorly placed and
badly handled in the *Dialogue*. It did *look* as if Galileo was
throwing a not too subtle insult in Urban's direction. And Urban
didn't forget it. Years later Galileo mentioned in a letter:

> I hear from Rome that His Eminence Cardinal Antonio [Barber-
> ini, the Pope's nephew] and the French Ambassador have spoken
> to His Holiness and attempted to convince him that I never
> had any intention of committing so sacrilegious an act as to
> make fun of His Holiness, as my malicious foes have persuaded
> him and which was the major cause of all my troubles.[55]

Still, Galileo had gone against the prohibition convinced that he
had the necessary proof. Unfortunately, the kind of evidence with
which he could have won the day simply was not available. Such
proof presupposed a new physics, the gravitational laws of Newton, the experiments of Foucault, the parallax observed in 1838
by Bessel.

Even in retrospect, the theological problem raised by the *Dialogue* was difficult. The faulty commitment of theologians to the
idea that texts from Sacred Scripture could solve strictly scientific
problems had been the cause of a decree which never should have
been issued. But once the prohibition was on the books and was
so openly violated, action of some kind seemed inevitable even to
those Church officials who favored Galileo.

By August, 1632, publication of the *Dialogue* was suspended
and unsold copies, if there were any, were confiscated. Three
weeks later, Galileo heard from Father Campanella:

[55] *Opere*, XVI, 455.

I have heard with great disgust that there is to be a commission of irate theologians to prohibit your *Dialogue*, and not a single person on this commission knows mathematics or other learned subjects. I caution you that, while you should hold that the opinion of the earth's motion was rightly prohibited, you are not held to believe that the contrary reasons are good ones.[56]

Campanella tried to get appointed to the special commission. That would at least give Galileo a chance. Unable to make the request personally, the Dominican Friar advised Galileo to have the Grand Duke suggest his name to the authorities in Rome. But it was too late. The commission had already been appointed and gone to work. Nor could Father Campanella be added to the board because, it was pointed out, he had defended Galileo in 1616 and was prejudiced in his favor. From a strictly legal point of view, of course, this was true. But it might well be asked how many of the judges who were chosen really were impartial. Unfortunately, Father Campanella, one of the most learned churchmen then living in Rome, was not even consulted on the case.

The special commission which had been formed to look into the Galileo matter summarized its finding in a report to the Pope on September 11, 1632. It listed eight counts, though only three main charges against the author of the *Dialogue*. First, he had transgressed orders by definitely defending the motion of the earth and the immobility of the sun. Second, he falsely attributed the motion of the tides to the earth's movement and the sun's stability. Third, he had been fraudulently silent about an absolute injunction served to him in 1616, "not to hold, teach, or defend in any way, verbally or in writing" his Copernican opinions. The report then listed several minor grievances against the *Dialogue*, but noted that if the book was thought to be valuable, it could be corrected and released.

The third charge came as a surprise even to the Pope. The injunction of February 26, 1616, if our view is correct, was served illegally upon Galileo by the Commissary-General of the Holy

[56] *Opere*, XIV, 373.

Office, in the presence of Cardinal Bellarmine and others. Bellarmine had instructed Galileo to ignore it, but it had been written into the records of the Holy Office. Now it was discovered and because no one in Rome was exactly sure what had taken place during the audience with Bellarmine more than sixteen years before, the written account seemed to be sufficient reason to take action against Galileo. The preliminary commission did not recommend such action. But there were those in Rome who not only recommended, but demanded that Galileo be punished.

CHAPTER VI

The Trial of Galileo

The findings of the Special Commission were turned over to the Holy Office and Galileo was summoned on October 1, 1632, to come to Rome. His health, which had failed generally with age, was very poor and he asked to be spared the trip. The answer was that he could come by easy stages, but he had better come. Galileo then sent a medical certificate which confirmed his weak condition. But Rome insisted, feeling that he was merely trying to delay or avoid the inevitable. The Grand Duke reluctantly advised Galileo to go. The political situation was such that there was nothing else he could do short of incurring an interdict and the Galileo business did not seem as serious as all that. The ailing scientist left for Rome on January 23, 1633, assisted by some of the Grand Duke's personal servants. He arrived at the Tuscan Embassy on February 13, and, in accord with the wishes of his royal host, was assured of comfort and care.

Galileo rested at the Villa Medici for nearly two months before he was moved to quarters in the building of the Holy Office. The concessions made to Galileo were unprecedented.[1] Contrary to the legendary picture of Galileo languishing in a dungeon, he was given a five-room suite and a servant to care for his needs. He did not spend a single minute in a prison cell.

On April 12, 1633, Galileo was brought to the Holy Office and the hearings began. A Dominican friar, Father Vincent Maculano da Firenzuola, as Commissary-General of the Holy Office, was in charge of prosecuting the case. The task must have been doubly hard on him, since he himself had admitted to Father Castelli in 1632, that he found the Copernican system quite acceptable and that the problem certainly could not be solved on the authority of Sacred Scripture.[2] But like Riccardi he had a job to do, and unlike Riccardi he was not going to bungle it.

The questioning began with Firenzuola asking Galileo whether or not he knew why he had been called before the Holy Office. He replied that he wasn't sure but it was probably because of his latest book. He was shown a copy of the *Dialogue* and admitted that he was the author. The Commissary-General then reviewed the events of 1616. He asked Galileo to recount the events surrounding the prohibition of Copernicus's *De revolutionibus*. The Florentine scientist answered that it had been decided that the Copernican opinion, if taken as an established fact, contradicted Holy Scripture and thus was permitted only as a hypothesis. This decision, he continued, was made known to him by Robert Cardinal Bellarmine. What, Firenzuola asked, had Cardinal Bellarmine told him? Galileo answered that the Cardinal informed him that the Copernican opinion, taken absolutely, must not be defended or held. Galileo then brought forth the certificate which the Cardinal had given him years before when he had asked for something in writing with which to protect his reputation. Firenzuola looked at the certificate and then asked if any other

---

[1] "There is no previous example of a person called before the Tribunal who has not even been held *in secrete*.... I know of no other, whether bishop, prelate, or noble, who has not been placed in custody upon arriving in Rome." Niccolini to Cioli, April 16, 1633, *Opere*, XV, 95.

[2] Castelli to Galileo, October 2, 1632, *Opere*, XIV, 401–402.

command had been given to him in 1616. Galileo said that he did not remember any other command. Did anyone command him not to "hold, defend, or teach in any way" the said opinion? As far as he could recall, Galileo answered, the only command which he had been given came from Cardinal Bellarmine and it was that he must not hold or defend the opinion. He did not remember the words "not to teach" or "in any way" but they might have been said. If they were, he had forgotten them. He had felt that there was no need to remember anything other than what was in Bellarmine's certificate, which he had saved all these years. Had he, the questioner went on, when seeking an *Imprimatur* for the *Dialogue*, revealed to the authorities that he had been given a command in 1616? Just what command the Commissary was referring to is not clear. Galileo was aware of only one instruction and that was from Cardinal Bellarmine. His answer was, at best, an evasion:

> I did not say anything about that command to the Master of the Palace when I asked for the *Imprimatur*. I did not think it necessary to say anything, because I had no doubts about it; for I have neither maintained nor defended in that book the opinion that the earth moves and that the sun is stationary, but have even demonstrated the opposite of the Copernican opinion and shown that the arguments of Copernicus are weak and inconclusive.[3]

That concluded the first hearing.

Five days later, on April 17, three theological Consultors of the Holy Office, who had been appointed to examine the text of the *Dialogue*, submitted their reports. They agreed that Galileo had taught the Copernican system. His book made it appear that a number of physical phenomena, among them the tides and the sunspots, which had alrealy been explained in other ways, could be caused only by a stable sun and a moving earth. He had also, in their opinion, defended the Copernican doctrine. Not only had he ridiculed Aristotle and Ptolemy, he had also proposed new arguments with the intention of establishing the truth of the for-

---

[3] *Opere*, XIX, 341.

bidden opinion. And he had held the Copernican doctrine. This, concluded the Consultors, was obvious in his work despite his testimony to the contrary. It was enough by way of evidence to point out that he mocked all who did not hold the Copernican theory. Surely Galileo did not consider himself numbered among the "mental pygmies" in which class he placed anyone who did not subscribe to the view of Copernicus.

The reports were filled with long quotations from the *Dialogue* confirming their conclusions. Historians have admitted freely that they were a just appraisal of Galileo's position.

Even so, there was not a very strong legal case against Galileo. In the first place, he did not remember any personal injunction having been given him by the Commissary-General in 1616. While this does not in any way mean that the injunction was planted in the files or that it was never really given, it does seem to support the position that the injunction had not been served validly. The fact that Galileo repeatedly appealed to Bellarmine's certificate might indicate that Bellarmine's explicit instructions to him were that he had only to refrain from holding or defending the Copernican system, no matter what anyone else, the Commissary-General included, might have said to him.

Secondly, it is true that he violated the intent, at least, of the decree of the Index dated March 5, 1616, which was directed against Copernican writings. Bellarmine's whole purpose in calling Galileo before him had been to make sure that he knew what the decree meant. In the words of the certificate he gave Galileo on May 26, 1616, it meant that the Copernican doctrine was contrary to Holy Scripture and therefore was not to be defended or held. This point must be emphasized, for, although it is true that the decree made a distinction between works which tried to reconcile the new astronomy with Holy Scripture and those which merely taught the system as fact, it is also true that, in silencing attempts to prove the reconcilability of the theory with Scripture, the decree was saying that the system was definitely contrary to the Bible. Therefore, not only the reconcilability, but the system itself, taken as a fact, was not to be defended or held. Galileo understood this. He admitted in his letter to Archduke Leopold of Austria that he held the Copernican system to be true "until it pleased the theologians to suspend the book [the *De revolutionibus*] and to declare the opinion to be false and repugant to Holy Scripture." Even though the Consultors' reports proved beyond doubt that he had violated the decree of the Index, Galileo still had a legal out. The Master of the Sacred Palace and the Inquisitors who gave him permission to print the *Dialogue* knew about that decree. If they thought that the book was not in accord with it, they should never have granted permission to publish.

What it all came down to was that the Pope, wishing, among other things, to have the name Barberini respected in the world of arts and sciences, had urged Galileo to write. Galileo might well have felt that Urban's support abrogated the instruction from Bellarmine years before, not to hold or defend the Copernican system. But then, in the *Dialogue*, he had ignored the Pope's instructions and gone too far. The Pope, convinced that he had been misused, withdrew his support and had Galileo judged according to the strict prohibitions of the questionable injunction and the decree of the Index. Partially, the conflict was Urban's fault. He did not read the *Dialogue* before publication, nor apparently did he make it clear to the censors exactly what could or could not be said in the book. Ciampoli and Riccardi must also share the responsibility: Ciampoli for pushing Riccardi to grant the *Imprimatur* and giving him a final clearance to do so, and Riccardi for actually allowing the work to be published even though he still had his doubts about its mode of presentation.

Yet there can be no doubt that Galileo took an untenable position when he claimed, as he did throughout the trial, that he had not defended or held the Copernican opinion. Some historians have expressed contempt for Galileo because he did not stand up to the Holy Office and defend his view, whatever the consequences. Such a view is wholly unrealistic. Galileo had no desire to be excommunicated from the Church. Besides, to the very end he kept alive a somewhat sanguine feeling that everything would somehow be all right. The rumors prevalent at the time of his arrival in Rome were that everything was going to be settled quietly. Galileo would be given a severe lecture on authority, and his book would be suspended until he recast it in a more hypothetical mold. Now all that looked like wishful thinking. The repeated questions about an absolute injunction which he did not remember and the lack of impression which his certificate from Bellarmine seemed to make on the authorities left Galileo less than secure at what might happen to him.

Still there was hope. It became evident that he was not without his defenders even on the ten-man board of cardinals assigned to judge the case. Francesco Cardinal Barberini, the Pope's own nephew, had little sympathy with his uncle's anger. When the prohibition had been forced in 1616, the issues had been more doctrinal than personal. But this time it was different. Undoubtedly some Church officials wanted to protect doctrinal authority and others still felt that Copernicanism was heretical. But the men who wanted to see Galileo subjected to the strictest penalties were not so highly motivated as all that. It is hard to say who they were or why they were so anxious to have Galileo grovel in the dust. But that there was a group of clerics bent on humiliating him cannot be denied. Cardinal Barberini, Firenzuola, Riccardi, and Campanella, among others, were pretty sure that this trial was not aimed so much at preserving doctrinal purity as it was at extracting personal revenge. And they wanted no part of it. No doubt they felt that some kind of personal penance was in order. But not complete humiliation.

As a result of what seems to have been a plan discussed by Firenzuola and Cardinal Barberini, a method was devised to spare Galileo as much as possible. Since Galileo had denied what was plainly evident in his book and added evasion to disobedience, the situation did not look hopeful. Firenzuola, with Cardinal Barberini's backing, obtained permission to deal extra-judicially with

Galileo: in other words, to make a deal with the accused. The idea was that if Galileo would tell the truth and admit that he had gone too far in his book, he might get off with a private penance and temporary house arrest, the *Dialogue* would be suspended at least until corrections were made, and the matter would end there. "In this way," Firenzuola wrote to Cardinal Barberini on April 28, "the court will maintain its reputation; it will be possible to deal leniently with the accused, and, whatever decision is reached, he will recognize the favor shown him." [4] Firenzuola's letter also describes how he convinced Galileo that accepting the deal was the wise thing to do:

> In order not to lose time, I talked to Galileo yesterday afternoon. After many arguments and counter-arguments, by the grace of God, I attained my objective and brought him to a full understanding of his error so that he clearly realized that he had erred and gone too far in his book. He expressed his deepest feelings with regard to all of this like one who felt great consolation in the recognition of his error. He was also willing to confess it judicially. He asked for a little time in order to consider the most fitting form for his confession, which, I hope, will be substantially as indicated. [5]

Two days later, on April 30, Galileo was called to appear for the second hearing. Asked if he wished to make a statement, he told the court:

> In the course of some days' continuous and intense reflection on the questions asked me on the sixteenth [Galileo is referring to the first hearing which was actually held on April 12] of the present month, and in particular as to whether, sixteen years ago, an injunction was intimated to me by order of the Holy Office, forbidding me to hold, defend, or teach in any way the opinion which had just been condemned . . . it occurred to me to reread my published *Dialogue*, which I had not seen for three years, in order carefully to ascertain whether, contrary to my most sincere intention, there had inadvertently fallen from my pen anything from which a reader or the authorities might infer not only some sign of disobedience on my part, but also anything

[4] *Opere*, XV, 107.
[5] *Ibid.*

else which would induce the belief that I had gone against the orders of the Holy Church.

Being, by the kind permission of the authorities, free to send my servant about, I was able to obtain a copy of my book. And, having obtained it, I applied myself most diligently to studying it and to considering it most carefully. And, because I had not seen it for so long, it seemed to me like a new writing and by another author. I freely confess that it seemed to me composed in such a way that a reader ignorant of my real purpose might have reason to think that the arguments presented for the false side, which I really intended to refute, were expressed in such a way as to be calculated rather to compel conviction by their soundness than to be easily solved.

Two arguments there are in particular—the one taken from the solar spots, the other from the ebb and flow of the tide— which really do strike the reader with a far greater show of force and power than should have been given to them by one who regarded them as inconclusive and who intended to refute them, as indeed I sincerely held and do hold them to be inconclusive and capable of being refuted. And as a personal excuse for having fallen into an error so foreign to my intention, not being content with merely saying that when someone gives the arguments for the opposite side with the object of refuting them, he should (especially if writing in the form of a dialogue) state these arguments in their strictest form and should not cover them over to the disadvantage of his opponents,—not being content, I say, with this excuse, I resorted to that of the natural complacency which every man feels with regard to his own subtleties and in showing himself more skillful than most men in devising, even in favor of false propositions, ingenious and plausible arguments. . . My error then, has been, and I confess it, one of vain-glorious ambition and of ignorance and inadvertence. . . .

Having completed his statement, he was dismissed. But he came back and added as an afterthought:

And in confirmation of my assertion that I have not held and do not hold as true the opinion which has been condemned, . . . if there shall be granted to me, as I desire, the means and the time to make a clearer demonstration of this, I am ready to do so; and there is a most favorable opportunity for this, for in the work already published the interlocutors agree to meet again after a certain time to discuss several distinct problems of Nature not connected with the matter already treated. As this affords me the opportunity of adding one or two other "Days" I promise to resume the arguments already brought in favor of the said opinion, which is false and has been condemned, and to refute them in such an effectual way as by the blessing of God may be supplied to me. I pray, therefore, this holy Tribunal to aid me in this good resolution and to enable me to put it in effect.[6]

Galileo admitted that he had gone too far in his book, but he persisted in denying that he had held the new system as true. Now he offered to publish an addition to his work which would not merely water down, but actually contradict his scientific beliefs. Fortunately, even the Holy Office felt no need for that. Koestler believes that

> ... his fears were exaggerated, and that his self-immolatory offer (which the Inquisition discreetly allowed to drop as if it had never been made) was quite unnecessary, is beside the point. His panic was due to psychological causes: it was the unavoidable reaction of one who thought himself capable of outwitting all and making a fool of the Pope himself, on suddenly discovering that he has been 'found out.' His belief in himself as a superman was shattered, his self-esteem punctured and deflated.[7]

I think that is an overstatement. It had never been his intention to make fun of the Pope. He had been bold, but only because there were reasons which seemed to indicate that he would be allowed some leeway. As Giorgio de Santillana has written:

> He knew, beyond doubt, that he had disregarded the Pope's explicit intentions. But he was strong in his certainty that he had not disobeyed the edicts of the Church. He obviously thought— as Ciampoli did for that matter—that he had to contend with the fancies of a vain and headstrong but brilliant personality who was still, for all his pontifical robes, the old Maffeo Barberini whom he had known and loved. There is scarcely an intelligent man who will not believe that people in power cannot take themselves seriously all of the time and who will not naïvely extend to them his sympathy in the hope of a shade of refreshing complicity in return, of a trace of humor.[8]

Now the law was being applied according to a strict interpreta-

---

[6] *Opere*, XIX, 342–344.
[7] Arthur Koestler, *The Sleepwalkers*, p. 489.
[8] G. de Santillana, *Crime of Galileo*, p. 203.

tion and Galileo knew he had lost his gamble and that some consequences would be forthcoming. He hid behind the denial that he had held or defended the theory, well aware that there was no way in which he could be forced to tell the whole truth.

After this hearing, Galileo was allowed to move back to the Tuscan Embassy at the Villa Medici, a privilege equivalent in our day to a defendant obtaining release without bond during the very process of his trial.

On May 10, Galileo was summoned before the court and asked if he wished to present a defense. Galileo handed Firenzuola a document which stated his side of the case. It is worth quoting in full:

> When asked if I had signified to the Reverend Father, the Master of the Sacred Palace, the injunction privately laid upon me, about sixteen years ago, by order of the Holy Office, not to hold, defend, or in any way teach the doctrine of the motion of the earth and the stability of the sun, I answered that I had not done so. And, not being questioned as to the reason why I had not intimated it, I had no opportunity to add anything further. It now appears to me necessary to state the reason in order to demonstrate the purity of my intention, ever foreign to the employment of simulation or deceit in any operation I engage in. I say, then, that as at that time reports were spread abroad by evil-disposed persons, to the effect that I had been summoned by the Lord Cardinal Bellarmine to abjure certain of my opinions and doctrines, and that I had consented to abjure them, and also to submit to punishment for them, I was thus constrained to apply to His Eminence, and to solicit him to furnish me with an attestation, explaining the cause for which I had been summoned before him: which attestation I obtained, in his own handwriting, and it is the same that I now produce with the present document. From this it clearly appears that it was merely announced to me that the doctrine attributed to Copernicus of the motion of the earth and the stability of the sun must not be held or defended, but that, beyond this general announcement affecting every one, no trace of any other injunction intimated to me appears there. Having, then, as a reminder, this authentic attestation in the handwriting of the very person who intimated the command to me, I made no further application of thought or memory with regard to the words employed in announcing to me the said order not to hold or defend the doctrine in question: so that the two

articles of the order—in addition to the injunction not to 'hold' or '*defend*' it—to wit, the words 'nor to teach it in any way whatsoever'—which I hear are contained in the order intimated to me, and registered—struck me as quite novel and as if I had not heard them before: and I do not think I ought to be disbelieved when I urge that in the course of fourteen or sixteen years I had lost all recollection of them, especially as I had no need to give any particular thought to them, having in my possession so authentic a reminder in writing. Now, if the said two articles be left out, and those two only retained which are noted in the accompanying attestation, there is no doubt that the injunction contained in the latter is the same command as that contained in the decree of the Sacred Congregation of the Index. Whence it appears to me that I have a reasonable excuse for not having notified the Master of the Sacred Palace of the command privately imposed upon me, it being the same as that of the Congregation of the Index.

Seeing also that my book was not subject to a stricter censorship than that made binding by the decree of the Index, it will, it appears to me, be sufficiently plain that I adopted the surest and most becoming method of having it guaranteed and purged of all shadow of taint, inasmuch as I handed it to the Supreme Inquisitor at the very time when many books dealing with the same matters were being prohibited solely in virtue of the said decree. After what I have now stated, I would confidently hope that the idea of my having knowingly and deliberately violated the command imposed upon me will henceforth be entirely banished from the minds of my most eminent and wise judges; so that those faults which are seen scattered throughout my book have not been artfully introduced with any concealed or other than sincere intention, but have only inadvertently fallen from my pen owing to a vainglorious ambition and complacency in desiring to appear more subtle than the generality of popular writers, as indeed in my other deposition I have confessed—which fault I shall be ready to correct by writing whenever I may be commanded or permitted by your Éminences.

Lastly, it remains for me to pray you to take into consideration my pitiable state of bodily indisposition, to which, at the age of seventy years, I have been reduced by ten months of constant mental anxiety and the fatigue of a long and toilsome journey at the most inclement season—together with the loss of the greater part of the years of which, from my previous condition of health, I had the prospect. I am persuaded and encouraged to do so by the clemency and goodness of the most eminent lords, my judges; with the hope that they may be pleased, in answer to my prayer, to remit what may appear to their entire justice to be lacking to such sufferings as adequate punishment—out of consideration for my declining age, which too, I humbly commend to them. And I would equally commend to their consideration my honor and reputation, aginst the calumnies of illwishers whose persistence in detracting from my good name may be inferred from the necessity which constrained me to procure from the Lord Cardinal Bellarmine the attestation which accompanies this.[9]

It was a strong defense. Since he did not remember any special command "not to teach or discuss in any way" he had not felt it necessary to tell the Master of the Sacred Palace about his audience with Cardinal Bellarmine. Father Riccardi should have known well the limitations which the decree of the Index imposed on astronomical writings. In other words, Galileo is asking why his book was officially approved and is now prohibited. That was a good question. On the other hand, the judges knew that he had knowingly gone against the provisions of Bellarmine's instruction, the absolute injunction and the decree of the Index, and that he had not been altogether truthful in his answers. He could not be dismissed without some salutary penance. According to the arrangement, all that remained now was to carry out the terms of Firenzuola's compromise and bring the case to a close.

At this point, a report summarizing the progress of the trial was drawn up for presentation to the Pope and the full Congregation of the Holy Office. It was a sloppy and misleading summary, obviously geared in favor of the prosecution. It contained statements from Galileo's *Letter to the Grand Duchess*, for example, which were drawn out of context. Thus the report cites Galileo as holding that "in Holy Scripture there are many propositions false as to the strict meaning of the words." The report then made it look like the absolute injunction was given to Galileo not by the Commissary-General, but by Bellarmine himself! In short, the report contained a whole list of ambiguous and misleading charges. It seems to have been the work of a "severist"

[9] Cited by F. S. Taylor, *Galileo and the Freedom of Thought*, pp. 156–159.

faction which was determined to teach Galileo, and anyone else who cared to "meddle" in the affairs of theology or authority, a hard lesson. Father Grienberger's remark that had he not incurred the anger of the Jesuits, Galileo could have written on the Copernican system for the rest of his life, and Father Riccardi's recorded statement that the Jesuits would fight him most bitterly, perhaps indicate that some members of the Society favored this severe policy.[10]

In any case the report was obviously not Firenzuola's work. De Santillana is probably right in thinking that the Proctor-Fiscal of the Holy Office, one Carlo Sinceri, and the Assessor, a Monsignor Paolo Febei were the key men in the anti-Galileo faction and that it was they who supervised the drawing of this highly slanted report.

Now the procedure was that the officials of the Holy Office who had been conducting the investigation turned their report in to the Pope and the ten Cardinal-judges of the Holy Office for a decision. It seems that the Pope, who had other things on his mind, and a majority of the judges, had nothing to go on but the misleading summary and numerous suggestions and proddings from the severist element. At any rate Firenzuola's implicit promise to Galileo that the trial would be allowed to die quietly was

[10] It would not be entirely accurate to blame the Jesuits for the condemnation of 1633 any more than it would be to hold the Dominicans responsible for the prohibition of 1616. Both times it was a matter of individuals acting, not an entire Society or Order. Both times there were Dominicans and Jesuits on each side of the question. De Santillana (*op. cit.*, p. 143) calls the Jesuits "trained seals" and claims that "obedience and a preoccupation with 'scandal' had been bred into the bones of the Order [The Jesuits are a Society, not an Order] from Bellarmine down, to such an extent that the intellectual reflexes were dead." The Professor holds the Jesuits responsible for the condemnation: "It was they and no one else, whose obligation it was to prevent the Pope from making a fool of himself. But the vast apparatus of indoctrination and constriction that their Order had devised, was now working to its own undoing. Following 'like unto corpses' the corporate political will of their Society, they shut their eyes, their ears and their minds. The power of discipline fed back into the complex steering machinery in a circuit of self-destruction" (*op. cit.*, p. 205). This blanket condemnation is almost as unfair and unrealistic as the description of the Jesuits which follows it.

overruled. On June 16, the decision of the Pope was entered into the records of the Holy Office. Instead of merely prohibiting the *Dialogue* "until corrected," giving Galileo a private penance and assigning him to house arrest, the scientist was to perform the humiliating act of formal abjuration and his book was to be forbidden:

> His Holiness decreed that the said Galileo is to be interrogated with regard to his intention, even with the threat of torture, and, if he sustains [that is, answers in a satisfactory manner), he is to abjure *de vehementi* (i.e., vehement suspicion of heresy] in a plenary assembly of the Congregation of the Holy Office, then is to be condemned to imprisonment as the Holy Congregation thinks best, and ordered not to treat further, in any way at all, either verbally or in writing, of the mobility of the earth and the stability of the sun; otherwise he will incur the penalties for relapse. The book entitled *Dialogo di Galileo Galilei Linceo* is to be prohibited. Furthermore, that these things may be known by all, he ordered that copies of the sentence be sent to all Apostolic Nuncios, to all Inquisitors against heretical pravity, and especially the Inquisitor in Florence, who shall read publicly the sentence in the presence of as many as possible of those who profess the mathematical art.[11]

Galileo had misused the Pope's friendship, openly disobeyed the express prohibitions against his theory, and plainly failed in his attempt to prove the Copernican astronomy. Still it is hard to see the logic of the sentence. There was no reason to condemn the *Dialogue* outright. It in no way attempted to reconcile the heliocentric system with Holy Scripture as Foscarini's book had done years before. The *Dialogue* defended the new astronomy as true on physical grounds. If this was objectionable, Galileo's book like Copernicus's *De revolutionibus*, should have been suspended until corrected and made more hypothetical. As for Galileo personally, the most that should have been given him was a penance for disobeying Bellarmine's admonition. Condemnation based on the absolute injunction was certainly tenuous, especially in the light of Bellarmine's certificate which was intended to state exactly what had happened to Galileo in 1616 and which made no men-

[11] *Opere*, XIX, p. 306–361.

tion of such an injunction. As for the decree of the Index, it was aimed at books, not authors.

The word "torture" in the document cited above has given rise to another legend connected with the name of Galileo, a legend which, until recently, enjoyed wide acceptance. Even today, Bertold Brecht's play on the life of Galileo misrepresents this phase of the famous condemnation. Galileo was not tortured, nor was he shown the instruments of torture. He was verbally threatened with torture. But he knew that the threat carried no weight. It was common knowledge, and canonists and moralists of the period are unanimous in verifying it, that no one of Galileo's age or poor health could be subjected to physical torture.[12] Now there can be no defense of torture, or even threats of torture. But the *territio verbalis* here was a mere formality and Galileo knew it.

On the morning of Tuesday, June 21, 1633, in accord with the stipulations of the papal decision, Galileo was questioned about his intention, that is, he was asked whether he really held the Copernican opinion. Had he, Firenzuola asked, or did he now believe that doctrine to be true? Galileo answered that since the decision of the Congregation of the Index had been revealed to him in 1616, he had held "as most true and indisputable, the opinion of Ptolemy, that is to say, the stability of the earth and the motion of the sun."[13] Very well. But, Firenzuola pointed out, your book seems to indicate otherwise. Galileo answered, ". . . I affirm therefore, on my conscience, that I do not now hold the condemned opinion and have not held it since the decision of the authorities."[14] But surely, at least when he wrote the book,

---

[12] "*Dicendum est cum Sanchez, Suarez, Narbona, etc., et Doctoribus in Tribunali Sancti Inquisitionis, diximus senem non esse torquendum esse senem 70 annorum, et etiam 60 si sit infirmae salutis, debilisque complexionis.*" Antoninus Diana, *Coordinati Resolutionum Moralium*, ed. Martin de Alcolea (Venice, 1728), V, Resol. XXIX, p. 337. This work was first published in 1658, but since Sanchez, Suarez, and Narbona all wrote well before 1633, the citation from Diana confirms that this was the common opinion at the time of the trial of Galileo.

[13] *Opere*, XIX, 361.

[14] *Ibid.*, 362.

he had held the Copernican opinion. Again he denied it. Finally, Galileo was told to speak the truth under threat of torture. He repeated that he had not held that opinion since 1616. There was nothing more that could be done, short of confronting him with a whole list of quotes from his *Dialogue* which seemed to contradict his repeated denials. That would have accomplished nothing. Torture was out of the question and Galileo knew it. He was sent back to his quarters to await sentence.

The next morning, Wednesday, June 22, 1633, Galileo was escorted to the Dominican convent of Santa Maria Sopra Minerva and the sentence was read to him. It began with a review of the whole case, starting with the first denunciation of his opinion to the Holy Office in 1615, then making public for the first time the verdict of the theological Consultors in 1616, on the two propositions which had been given them to examine, then recalling the audience with Cardinal Bellarmine and the injunction given to Galileo by the Commissary-General in 1616, and the decree of the Index which had declared the opinion to be contrary to Holy Scripture. It cited the fact that Galileo had written the *Dialogue* in support of the forbidden position despite his protestations to the contrary. And even if he did not remember the personal injunctions, he had violated the admonition of Cardinal Bellarmine not to hold or defend the theory. Then it gave the sentence:

> We say, pronounce, sentence, and declare that you, the said Galileo, by reasons of the matters brought forth in trial, and by you confessed as above, have rendered yourself in the judgment of this Holy Office vehemently suspected of heresy, namely, of having believed and held the doctrine which is false and contrary to the Sacred and Divine Scriptures, that the sun is the center of the world and does not move from east to west and that the earth moves and is not the center of the world; and that any opinion may be held and defended as probable after it has been declared and defined contrary to Holy Scripture; and that consequently you have incurred all the censures and penalties imposed and promulgated in the sacred canons and other constitutions, general and particular, against such delinquents. From which

we are content that you be absolved, provided that, first, with a sincere heart and unfeigned faith, you abjure, curse, and detest before us the aforesaid errors and heresies and every other error and heresy contrary to the Catholic and Apostolic Roman Church in the form to be prescribed by us for you.

And, in order that your grave and pernicious error and transgression may not remain altogether unpunished and that you may be more cautious in the future and an example to others that they may abstain from similar delinquencies, we ordain that the book *Dialogue of Galileo Galilei* be prohibited by public edict.

We condemn you to the formal prison of the Holy Office subject to our judgment, and by way of salutary penance we prescribe that for three years to come, you repeat once a week the seven penitential psalms. Reserving to ourselves the right to moderate, commute, or take off, in whole or in part, the aforesaid penalties and penance.

And so we say, pronounce, sentence, declare, ordain, and reserve in this and in any other better way and form which we can and may rightfully employ.[15]

The decree of Sentence was signed by only seven of the ten cardinal-judges of the Holy Office. Conspicuously absent was the signature of Francesco Cardinal Barberini, who had advocated leniency throughout the proceedings and probably considered the decision to be of questionable justice and needless severity. Cardinals Borgia and Zacchia also did not sign the decree, but their motives for abstention are not known.

Galileo, having heard the sentence, expressed a willingness to abjure as required, but asked that two charges be omitted from the formula of abjuration. One was a statement which hinted that he was not a good Catholic; the other was an inference that he had obtained the *Imprimatur* by devious or cunning methods. The officials had no desire to argue the matter and they granted his request. Galileo knelt to recite his formal abjuration:

I, Galileo, son of the late Vincenzio Galilei, Florentine, aged seventy years, arraigned personally before this tribunal and kneeling before you, Most Eminent and Lord Cardinals Inquisitors-General against heretical pravity throughout the entire Christian

[15] *Ibid.*, 405–406.

Commonwealth, having before my eyes and touching with my hands the Holy Gospels, swear that I have always believed, do believe, and with God's help will in the future believe all that is held, preached and taught by the Holy Catholic and Apostolic Church. But, whereas, after an injunction had been lawfully intimated to me by this Holy Office to the effect that I must altogether abandon the false opinion that the sun is the center of the world and immobile, and that the earth is not the center of the world and moves, and that I must not hold, defend, or teach, in any way, verbally or in writing, the said false doctrine, and after it had been notified to me that the said doctrine was contrary to Holy Scripture, I wrote and printed a book in which I treated this new doctrine already condemned and brought forth arguments in its favor without presenting any solution for them, I have been judged to be vehemently suspected of heresy, that is, of having held and believed that the sun is the center of the world and immobile and that the earth is not the center and moves.

Therefore, desiring to remove from the minds of Your Eminences, and of all faithful Christians, this vehement suspicion rightly conceived against me, with sincere heart and unpretended faith I abjure, curse, and detest the aforesaid errors and heresies and also every other error, error and sect whatever, contrary to the Holy Church, and I swear that in the future I will never again say or assert verbally or in writing, anything that might cause a similar suspicion toward me; further, should I know any heretic or person suspected of heresy, I will denounce him to this Holy Office or to the Inquisitor or Ordinary of the place where I may be.

Further, I swear and promise to carry out and observe in their integrity, all penances that have been or shall be imposed upon me by this Holy Office. And if I should violate, which God forbid, any of these my promises and oaths, I submit myself to all the castigations and penalties imposed and promulgated in the sacred canons and other constitutions, general and particular, against such delinquents. So help me God and these Holy Gospels which I touch with my hands.[16]

The condemnation of Galileo was now complete. The scientist had tried to batter down the old view of the universe and the traditional exegesis of Scripture by beating his head against a wall

---

[16] *Ibid.*, 406. There is a myth that Galileo, upon leaving the Minerva, looked up at the sky and, stamping his foot cried, *"Eppur si muove"*—"Yet it does move!" This particular bit of apocrypha was started by Guiseppe Baretti in 1757 and has been widely repeated in uncritical writings through the years.

of conservatism and mocking those who felt that it should not be torn down. The wall stood; Galileo's tools had not been the best, nor had he used them effectively. It was his own admirable but imprudent doggedness as much as the strength of the wall that finally drove him to his knees.

Galileo was sentenced as vehemently suspected of heresy. This was an unfortunate decision on several accounts. First, the Copernican opinion was treated as heretical when, in reality, it was not. The eleven theological Consultors in 1616 had qualified the proposition that the earth moves as "formally heretical" but their judgment did not make the immobility of the earth a matter of faith. Neither the ordinary nor the extraordinary magisterium of the Church had pronounced infallibly on the Copernican system. St. Augustine and St. Thomas Aquinas had said long before that Scripture was not meant to teach natural science. In other words, Galileo was suspected of heresy for holding something which it was not heretical to hold, since the Copernican opinion had never been declared false by infallible authority.

Second, the Congregation of the Index had decreed that the Copernican position was not to be defended or held, and external obedience was required even if Galileo had sufficient reason to withhold internal assent with regard to the doctrine supported by the decree. They also believed that he had been given an injunction not to discuss the matter in any way. Thus the Holy Office would have been on better legal footing had it convicted him of disobedience rather than of suspected heresy.

Third, the fact that Galileo was made to abjure his opinion supplies the emotional content necessary to make the Galileo story apt material for polemics against the Church. Had Firenzuola's suggestions been carried out, it is highly doubtful that the famous conflict would have had such a long-lasting and widespread prominence. Had Galileo been given a private penance and told not to write any more on the matter, history's judgment would have been less severe on his judges; they were, after all, unprepared for the new science. But the fact that they made him publicly disown his conviction accentuated their error and confirmed a tragic mistake which has never been forgiven.

The decree of Sentence was both doctrinal and disciplinary. Like the decree of the Index seventeen years before, it was approved by the Pope in the common way (*forma communi*) and thus remained only a fallible act of a Sacred Congregation in no way involving an infallible commitment of authority. Those who have tried to use it to attack the infallibility of the Pope show complete misunderstanding of the Church's doctrine on papal infallibility and overlook the fact that even in the seventeenth century, Catholic philosophers and theologians realized that the decree of the Holy Office did not make the immobility of the earth or the mobility of the sun a matter of faith. Descartes wrote to Mersenne in 1634, "As I do not see that this censure has been confirmed either by a Council or by the Pope, but proceeds solely from a committee of cardinals, it may still happen to the Copernican theory as it did to that of the antipodes which was once condemned in the same way." [17] The Jesuit Father Riccioli wrote in 1651:

> Inasmuch as no dogmatic decision was rendered in this case, either on the part of the Pope or on the part of a Council ruled by the Pope and approved by him, it is not, by virtue of that decree of the Congregation, a doctrine of faith that the sun is moving and the earth standing still. . . . Yet every Catholic is bound by virtue of obedience to conform to the decree of the Congregation, or at least not to teach what is directly opposed to it. [18]

In 1676 the moral theologian John Caramuel noted that:

[17] *Opere*, XVI, 89. The reference to the Antipodes goes back to the eighth century and a letter from Pope Zacharias to St. Boniface concerning the doctrine of a priest named Vigilius, who, it seems, held that there existed "another world under the earth inhabited by people." Zacharias's letter merely stated that Vigilius *should* be condemned for such a doctrine. But, as far as we can discern, the condemnation never took place. In fact, Vigilius was soon appointed Bishop of Salzburg. As Descartes predicted, the prohibition against Copernican writings was revoked. In 1758, the general ban against Copernican works was abrogated. The special prohibitions were repealed in 1822 and the next edition of the Index of Forbidden Books, published in 1835, contained no trace of the infamous condemnation of heliocentric astronomy.

[18] Riccioli, *Almagestum Novum* (Bologna: 1651), p. 162.

The cardinals by their declaration (of 1633) did not establish the motion of the earth as a heresy . . . when an opinion is condemned by the cardinals, it is condemned as a practical consideration. A proposition thus condemned is not changed into heresy, but it loses all extrinsic authority and is rendered for all practical purposes, improbable . . .[19]

After the trial and abjuration, the sentence was commuted. Galileo received permission to have his daughter, Sister Maria Celeste, a Carmelite nun, recite the penitential psalms in his stead. There was to be no "formal prison of the Holy Office" either. On June 30, 1633, Galileo was released in the custody of his friend Archbishop Ascanio Piccolomini of Siena. The Archbishop placed his palace and staff at Galileo's disposal, encouraging Galileo to rest and regain his spirit. Galileo remained in Siena until December, when he was allowed to move back to his country estate at Arcetri, near Florence. Forbidden to write any more on the Copernican astronomy, Galileo began systematically arranging his ideas for a new physics.

In 1635, the first Latin translation of the *Dialogue* was published by Bernegger at Strasbourg, out of reach of the Holy Office. The following year saw the completion of Galileo's *Discourses concerning Two New Sciences*. It was undoubtedly his greatest contribution to modern science. The controlled thought experiment, the mathematical conception of nature and motion so forcefully presented in this book, set a pattern which, with necessary modifications, Newton was to follow with such great success. The *Discorsi* was an astounding intellectual accomplishment for any man, especially one who was seventy-two years of age. On this

[19] John Caramuel, *Theologia Fundamentalis* (Lyons: 1676), I pp. 104–106. H. Grisar (*Galileistudien*: pp. 337 ff) has shown that though Copernican writings remained officially on the Index until 1822, this did very little harm to astronomical research and development. The works of Copernicus and Galileo were always able to be read by anyone who took the trouble to obtain permission. Furthermore, it was always allowed at any time to write on the heliocentric system as a hypothesis. The Jesuit Father Kochansky wrote in 1685 that any Catholic was free "to look for an irrefutable mathematical and physical demonstration of the motion of the earth." See J. Donat, *The Freedom of Science* (New York: Wagner, 1914, pp. 196 ff.

work Galileo's true fame would have rested had it not been for the famous condemnation.

In 1637 Galileo suffered the loss of sight in both eyes. His courage, so often tested during his lifetime, was able to meet and overcome even this new obstacle. He moved into the city of Florence and continued to work. He spent hours each day dictating additional chapters for his *Discorsi*, teaching and encouraging those who came to him for instruction. In the following year, his *Discourses concerning Two New Sciences* was published at Leyden.

On January 8, 1642, the year in which Isaac Newton was born, Galileo died. The great scientist was buried in the Church of Santa Croce in Florence. Later, his remains were moved next to the tombs of two other famous Florentines, Michaelangelo and Machiavelli, in the same Church. When the Grand Duke petitioned for permission to place a monument over the tomb of Galileo, he was refused and told that this would not be fitting since Galileo had given rise to the "greatest scandal in Christendom." In a sense, the tragic conflict between Galileo and his Church merits that title. For centuries, Galileo's name has been invoked against papal infallibility and to "prove" that the Roman Catholic Church is a vicious enemy of science and progress. These claims can be refuted by an objective study of the facts. But the refutations have never been as attractive as the accusations to that part of the human temperament which more readily believes scandal than fact. But there is no doubt that the condemnation was the result of an unnecessary conflict, the *entire* circumstances of which we may never know.

CHAPTER VII

The Galileo Case Today

In any treatment of the tragic condemnation, perhaps the most important single factor, the one that will determine in large measure whether or not a true historical account is given, is the writer's particular insight into the character and personality of Galileo himself and into the psychological effects of his discoveries on society as a whole. Galileo was not infallible: not a great errorless crusader. Nor was he a conceited, obstinate troublemaker. He was a man of scientific genius, an extrovert, and an optimist. His temperament shows through in his writings; he was independent, quick-witted, scornful of his opposition. He did not think himself greater than he was, but neither was he one to underestimate his own ability, as is seen in this passage from the *Assayer*:

> Others, not wanting to agree with my ideas, advanced ridiculous and impossible opinions against me; and some, overwhelmed and convinced by my arguments, attempted to rob me of that glory which was mine, pretending not to have seen my writings and trying to represent themselves as the original discoverers of these impressive marvels.[1]

[1] Galileo, *The Assayer*, trans. S. Drake, *Discoveries and Opinions of Galileo*, p. 232.

And in the *Dialogue*:

> Harken then to this great and new wonder. The first discoverer of the solar spots, as also of all other celestial novelties, was our Lincean academician [Galileo] and he discovered them in the year 1610.[2]

Galileo's temperament was such that once he had made discoveries with his telescope, which were new and important, he was convinced that a new age had begun and that he was, as Campanella described him, "another Columbus," disproving man's false beliefs and leading the way to progress and development. He had succeeded where all humanity before him had failed. The desire to communicate his conviction and convince others, especially the Church authorities, that man's whole world view needed to be awakened from its sleepy stability, drove him to great lengths. Still, he was a man of his times, and whether he liked it or not, he should at least have recognized the fact that mankind was not about to accept, on his word alone, an entirely new view of the universe; especially one that removed the world from its comfortable place in the center of the heavens and made of it just another planet. Man liked to think that all creation revolved around him, that the heavens surrounding him were put there, just as flowers are planted in a park, for his enjoyment.

Too much was against him. Not just physics, not merely fundamentalist exegesis, but centuries of acceptance by everyone or nearly everyone, that the sun really moved and the earth did not. Men of all ages took it for granted that that was the way God had established the heavens. When Copernicus first suggested that the earth was in motion around the sun, he was publicly mocked. But when Galileo with the aid of his telescope tried to show that Copernicus was right, the attitude of the opposition passed from scorn to hostility. From the time of Galileo's discoveries in 1610 until the decision of the theologians in 1616, only six years had intervened. To ask the abandonment of philosophical, exegetical and human tradition in such a short time was, in the circum-

---

[2] Galileo, *Dialogue on the Great World Systems*, trans. de Santillana, p. 354.

stances, asking too much. Yet that was precisely what Galileo asked—or demanded. He would settle for nothing less.

By challenging the traditional view of the universe, Galileo upset the psychological security that derived from the neatly ordered inter-related hierarchies of astronomy, philosophy and theology. His opponents were afraid that if the geocentric concept fell, the whole construct of cosmology, the truth of the Scriptures and the anthropocentrism of creation would have to fall with it. Galileo showed no sympathy for the sensitivity of hierarchic relations. He plunged ahead with a process of demythologizing that even on a non-religious, purely human level, caused a psychological vacuum. Like nature, the human psyche abhors a vacuum. The solution is not that Galileo should have kept silence. But a more humble and realistic man might have proceeded with a better understanding of the monumental implications of the new astronomy.

Galileo won the final victory, but with byproducts quite foreign to his intention. He would not have wanted his name to be used against the authority and infallibility of his Church. The one thing he probably did wish posterity to learn from his personal tragedy was that no new opinion is wrong simply because it is new. That lesson alone would be a worthy legacy.

As I explained at the beginning of this book, my intention in writing it has been to present the facts of the case especially in the light of the scientific, philosophical and theological issues involved, and thereby to challenge the biased, or at least unhistorical, treatment that the conflict has received so frequently through the years. Undoubtedly the major fault of most writings on the subject is oversimplification. To see the condemnation as the logical move of a power-hungry Church is as false as to say that Galileo was wholly to blame because he would not confine his statements to the logical limitations of his evidence.

Catholics will always be in the unfortunate position of having to admit that a court of Catholic theologians condemned a doctrine and a man who, as it turned out, were right. In addition, they will have to regret that it was not until 1822 that the prohibition against Copernican works was removed from the Index and that only at the Second Vatican Council was the condemnation decried by Church officials as "something that is not altogether without fault on our part" and as "miserable and unjust." But this does not mean that they are required to accept in silence abuse hurled at the Church in the name of Galileo. It is important to the future of the entire human quest for knowledge that the Church's position on progress and science be understood correctly and not distorted by half-truths, inferences and insults. Catholics should be concerned when they read in recently published or reprinted books:

> Galileo popularized and rigorously demonstrated the Corpernican system.... In 1616 the Roman Inquisition condemned for the first time the opinion of Copernicus as contrary to the Bible and reason. Later, Galileo was also condemned, jailed, subjected to moral torture, threatened with physical torture, and sequestered during the rest of his days.[3]

> There were intrigues and counter-intrigues, plots and counter-plots, lying and spying; and in the thickest of this seething, squabbling, screaming mass of priests, bishops, archbishops and cardinals, appear two popes, Paul V and Urban VIII.[4]

A New York City lawyer, in a book entitled *Man on Trial*, simply asserts the naïve view that the proceedings against Galileo constituted "the climax of the onslaught of organized religion against scientific progress." [5] What it all amounts to is nothing more than the same old charges being passed off as uncontested facts.

However, it is encouraging to note that with the critical study of the history of science now being carried on, a more balanced view of the condemnation is emerging. The Galileo story itself

---

[3] A. Fouille, *Historia general de la filosofia* (Santiago de Chile, 1955), pp. 274 f.

[4] Andrew D. White, *A History of the Warfare of Science with Theology in Christendom*, excerpt reprinted in *The Achievement of Galileo*, ed. J. Brophy and H. Paolucci (New York: Twayne, 1962), pp. 195f.

[5] Gerald Dickler, *Man on Trial* (Garden City: Doubleday, 1962), p. 61.

is being demythologized. Such outstanding authorities in the history of science as E. A. Burtt, I. B. Cohen, A. C. Crombie, E. J. Dijksterhuis, and Stillman Drake, among others, while not overlooking the fact that the Church officials committed a lamentable error, assert the complexity of the situation in its historical and doctrinal context. Burtt points out that:

> We are so accustomed to think of the opposition to the great astronomer [Copernicus] as being founded primarily on theological considerations (which was, of course, largely true at the time) that we are apt to forget the solid scientific objections that could have been, and were, urged against it.[6]

A recent book on freedom expresses a reasonable view of the case:

> No less obviously, this emancipation had been achieved against the opposition of the Church, so it brings us to the too familiar, tiresome theme of the conflict between science and religion. We now realize, or ought to, that the "victory" of science was not won by a glorious, hard-fought campaign, nor was it simply a victory of right over wrong. The early champions of science were defending ideas only half true, or right for the wrong reasons; the Church had good logical reason to oppose them apart from its concern for the faith. In fairness to both, we may conclude that there is no necessary conflict between science and religion rightly understood.[7]

I. B. Cohen notes that Galileo must have had a special sense of urgency to convert his fellow men to the Copernican system. In fact his conflict arose because he was a true believer and could see no way to have separate secular and theological cosmologies. Thus he fought with every weapon in his arsenal to make the Church accept a new system of the universe.[8]

One could cite passages from all the authors named above to

[6] E. A. Burtt, *The Metaphysical Foundations of Modern Science* (Garden City: Doubleday Anchor, 1955), p. 36.
[7] H. J. Muller, *Freedom in the Western World* (New York: Harper and Row, 1963), p. 254.
[8] I. B. Cohen, *The Birth of a New Physics* (Garden City: Doubleday, Anchor, 1960), p. 128.

the effect that the Galileo case is far more complex and the mistakes are far more understandable than most writings on it indicate. I am by no means trying to insist that everyone have his say and be done with it, as long as he treats the Church politely. But it is time that we separate fact from fiction and see the case, not in the light of how much blame each participant deserves, but in terms of what the problems were and where the mistakes were made. Clearly, the doctrinal issues raised by Galileo and those who opposed him are, with minor modifications, still crucial problems today. The rise of science and its divorce from philosophy, the relations between faith and science, the position of the Catholic Church relative to science and freedom: these were prominent difficulties in the Galileo case and it is important that in this book on Galileo we examine the same problems in their contemporary context.

PHILOSOPHY AND SCIENCE

In order to understand the relationship between post-Galilean science and philosophy, we should recall that before the successful launching of the new physics in the seventeenth century, science, as we generally use that term today, was considered to be the detailed, empirical extension of the general philosophy of nature. This structuring of a broad philosophy of nature which was to embrace, interpret and order the findings of its particular branches (e.g., biology, psychology, geology, etc.) goes back to Aristotle and his scholastic followers. To them the term "philosophy" was equated with the quest for the fullness of natural, human knowledge and wisdom attainable in this life. Since, according to strong strains in the developed Aristotelian tradition, human knowledge could be distinguished according to three distinct degrees of abstraction, speculative philosophy was divided into three main and autonomous branches: the philosophy of nature, mathematics, and metaphysics.

According to this view, in the first degree of abstraction, the intellect concentrates on universal sensible matter, abstracted from individual sensible objects existing in extra-mental reality. When we observe several trees or several stones, we get a generalized concept of tree or stone. The individual notes of any particular tree or stone are left behind. By looking at physical reality in this way, in terms of universal sensible matter as it undergoes physical changes, the mind can attain a scientific knowledge of nature and the natural world. This is the domain of the philosophy of nature. The general philosophy of nature provides the principles and framework for further, specialized investigation in the particular sciences of nature. In other words, the principles of natural philosophy would be extended and applied in geology, biology, anthropology, psychology, physiology, and cosmology. The findings of these special sciences would always be related to the general philosophy of nature and thus fit into the picture as particularizing elements of an all-embracing view of physical reality.

In the second degree of abstraction, the intellect leaves behind the universal sensible matter studied in the philosophy of nature and concentrates on matter precisely as quantified. This level of abstraction is proper to the science of mathematics. Because mathematics abstracts from sensible qualities so important in the study of nature, the Aristotelians minimized its importance in the study of physical reality.

In the third degree of abstraction, the mind relinquishes any reference to matter at all. This is the level of abstraction proper to metaphysics. The subject matter of metaphysics is everything that exists, whether physical or spiritual, precisely under the aspect of its existence or being. The metaphysician studies the general notion of cause, property, principle, potency, act, truth, goodness, unity, beauty, etc., as part of his task of explaining being as being. In seeking the first cause of being, the metaphysician comes to a human knowledge of God and can reason to some of God's attributes, such as omnipotence, eternity, immutability and unicity. That part of metaphysics that studies God is called natural theology or theodicy.

The Aristotelians, then, divided philosophy as follows:

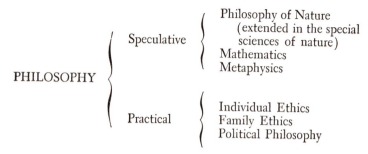

PHILOSOPHY

Speculative
- Philosophy of Nature (extended in the special sciences of nature)
- Mathematics
- Metaphysics

Practical
- Individual Ethics
- Family Ethics
- Political Philosophy

The Aristotelians considered mathematical physics to be a "middle science" between mathematics and the philosophy of nature. This science studies physical, sensible matter, but by means of applied mathematics. It does not give a complete picture of nature. In technical terms, mathematical physics contributes to the understanding of material and formal causes but not of efficient and final causes of a physical body or its motion. This is important because in the Aristotelian system, the final cause, the goal toward which a natural motion tends, was called the "cause of causes."

With the fall of the old astronomy, the entire structure and content of natural philosophy as understood by Aristotle was discredited. The traditional view of a unified philosophical-scientific approach to nature was discarded and the term "science" began to take on a new meaning.

With Galileo's announcement that the book of nature is written *only* in mathematical characters the march away from Aristotle began in earnest. Obviously if only geometrical demonstrations could describe the wonders of nature, a philosophy of nature which attempted to discover the inner nature of things by observing their secondary sense qualities, was little more than a dream. Galileo's mathematical realism led him to say that only the primary qualities: number, figure, size, position and motion, were real qualities in the objective world. Secondary qualities: tastes, odors, colors etc., were mere names for purely subjective phenomena. A gap was opened up between the knower and the known and consequently between the traditional philosophy and the new science. Galileo was interested in the measurement, not the goal, of nature. Nature was no longer directed toward man as in the scholastic view, but was given an independent existence. As Burtt has said:

> The natural world was now portrayed as a vast, self-contained, mathematical machine, consisting of motions of matter in space and time, and man with his purposes, feelings, and secondary qualities was shoved apart as an unimportant spectator and semi-real effect of the great mathematical drama outside.[9]

Galileo who had done so much to tear down the old universe contributed the spirit and the method that was to help in the construction of the World Machine.

René Descartes (1596–1650) was no less vehement than Galileo in asserting that mathematics alone could penetrate the secrets of nature. He saw nature as a gigantic geometric design that could be reduced to algebraic formulae in which there was no room for secondary qualities. With Descartes the division between knower and known, between mind and matter, is total and unmistakable. The world of extended bodies, the cosmic machine, exists with or without man. Man, the *res cogitans*, has an inner realm that is non-extended or spiritual, a soul imprisoned in the body. Even though Descartes exempted the soul from the sweep of his materialism, he did not work out a satisfactory theory of how the mind and body of man interacted. They met, he said, at the pineal gland. Apparently man should have been grateful for the concession of at least some kind of soul. The rest of vital nature, plants and animals, were denied sensitive souls and conceived as machines whose motions can be traced to mechanical processes. As one historian has described it, Descartes saw nature as "a machine and nothing but a machine; purposes and spiritual significance alike had been banished. . . . Intoxicated by his vision

[9] E. A. Burtt, *op. cit.*, p. 104.

and his success, he boasted, 'Give me extension and motion and I will construct the universe.' " [10]

With Thomas Hobbes (1588–1679), philosophical materialism reached its classic formulation. Whatever exists is matter and whatever changes is motion. It is as simple as that. Not even the mind of man can be saved from the flood of matter in motion. Knowledge is caused by the external object pressing on the sense organ and evoking a motive response from the brain. The right kind of sense stimulus will produce the right kind of "knowledge-response." Man has become an object in a fully objectified world.

It was left to Sir Isaac Newton (1642–1727) to render intelligible the "infinite and monotonous mathematical machine." By discovering the inverse-square law of gravitation, adding a *vis iner-tiae* to bodies, and conceiving of space and time as absolute frames of reference, Newton provided the new science with cogent and compelling accuracy. After observation and calculation, Newton's laws could be invoked and predictions made which were always or nearly always verified by further observation. Everything seemed to fit into place. The greatness of Newton's synthesis, which combined insights of Galileo, Kepler, Descartes, Borelli and others, was that it allowed science to build on a uniform, universally applicable concept of motion.

Newton accepted the world of matter as a world of atoms with mathematical properties. Since action at a distance was impossible, there also had to exist a medium, some kind of ether which could propagate motion across distances and vibrate when necessary so that electric and magnetic phenomena could be accounted for conveniently. Newton was enough of an empiricist to acknowledge that there might be more to matter than the mind can measure. He could not, for example, discover the cause of gravity from measuring phenomena, so he simply said that he "framed no hypotheses." But while the master allowed for the possibility that reality might be much more than matter in mo-

[10] J. H. Randall, *The Making of the Modern Mind* (New York: Houghton Mifflin, 1940), p. 259.

tion, his disciples did not. Absolute Space, Absolute Time, and absolutely infallible laws of nature added up to only one valid approach to the sensible world, and that approach was mathematical. It mattered little that man was now a mere observer of the cosmic automaton in action. Even if the human soul was reduced to some kind of special entity locked in that part of the brain called by Newton the *sensorium,* man could still solve the riddle of the universal machine's rhythm and thereby unlock the secrets of nature.

The logical implications of the whole mathematical-physics metaphysic created by classical science were drawn by Pierre Simon de Laplace (1749–1827). One day, he declared triumphantly, man would be able to measure exactly and predict infallibly the movement of the largest bodies and the lightest atoms in the universe. On that day, the future, like the past, would be present to the mind of man. The old struggle futilely pursued by philosophy to understand the nature of man and the cosmos was finally over. A new day had dawned. The philosophy of scientism had redeemed man from his faltering ignorance. Alexander Pope expressed the spirit of the eighteenth-century intellectuals:

Nature and Nature's laws were hid in night;
God said, Let Newton be! And all was light.

E. A. Burtt has given a vivid description of the transformation brought about by the new and reigning world-view. It is worth quoting in full:

> The gloriously romantic universe of Dante and Milton, that set no bounds to the imagination of man as it played over space and time, had now been swept away. Space was identified with the realm of geometry, time with the continuity of number. The world that people had thought themselves living in—a world rich with color and sound, redolent with fragrance, filled with gladness, love and beauty, speaking everywhere of purposive harmony and creative ideals—was crowded now into minute corners in the brains of scattered organic beings. The really important world outside was a world hard, cold, colorless, silent and dead; a world of quantity, a world of mathematically computable motions in mechanical regularity. The world of qualities as immediately per-

ceived by man became just a curious and quite minor effect of that infinite machine beyond.[11]

The logical limitations of science and the scientific method were trampled down in the rush to apply Newtonian mechanism to nearly every important area of study and human concern.[12] Although William Harvey's demonstration of the circulation of blood was Aristotelian in method and inspiration, it prompted theories as to how the laws of mechanics applied to living organisms. Some saw biology as a branch of the science of mechanics. Others proceeded on the principle that the human mind, like nature, must be governed by the law of universal gravity. Saint-Simon in France and Jeremy Bentham in England envisioned human society itself organized on the pattern of the new science: perfectly ordered and scientifically managed. In economics, Joseph Townsend devised a system of economic balance founded on natural forces and claimed it would work regardless of human effort or intervention. In such a climate it was only natural that the mechanists "chase from their company the philosophers, moralists, and metaphysicians, just as the astronomers and chemists have chased out the alchemists." [13]

An attempt was made by Immanuel Kant (1724–1804) to accommodate both the Newtonian physics and the power of human reason. His solution, though intricate and ingenious, kept nature and mind divided by an epistemological fence. On one side were the ideas of reason, on the other the observations of empirical science. There might have been activity on either side of the fence, but there was no continuity, no flow from one side to the other. On one side were the noumena (things-in-themselves) and on the other, the phenomena (the external appearances of things). Only the phenomena could be known. This meant that there could be no knowledge of things in themselves, of essences, of natures, and,

[11] E. A. Burtt, *op. cit.*, pp. 238f.
[12] See Floyd Matson, *The Broken Image* (New York: Braziller, 1964). This work, to which I am heavily indebted, is an excellent synthesis of the rise and fall of mechanism in physics and the natural and social sciences.
[13] Saint-Simon, cited by Matson, *op. cit.*, p. 32.

as a consequence, both the philosopher and the scientist were left agnostic as to the ultimate reality. From Kant's time onward, it became

> almost axiomatic that we can have true knowledge only of empiric objects, never of things lying behind the experience of the senses; our ideas are merely subjective constructions of the reason, which obtain weight and meaning only by applying them to objects of sense experiment. Hence God, immortality, freedom, and the like, remain forever outside the field of our theoretical or cognitive reason.[14]

In the nineteenth century the image of man as the neutral observer of objects, the slave of his environment, the robot moved by the forces of nature, was projected into even sharper focus following the publication of Darwin's evolutionary theory. As far as the mechanists were concerned, this was just what was needed to clear away the remnants of human will, spiritual conceptions of human destiny and whatever else claimed exemption from the automatic control of matter in motion according to determined laws. The fact that purely material evolution was a contradiction in terms and that evolution argued for, rather than against, purpose in nature did not bother advocates of the great Cosmic Machine. Juxtapose the atoms, apply the laws, and out comes man. And the man produced can be molded to fit any desired behavior pattern. Give the right stimulus and there will be the right response. On this principle, societies could be formed, stimulated, and controlled. All it cost was the humanity of man.

The time from Galileo until the end of the nineteenth century witnessed a complete revolution. The parallels in the intellectual climate at the beginning and end of this period are enlightening. No longer was it the geocentric system that could not be challenged, now it was the universal machine with all its running parts. No longer was it deemed necessary to defend man's privileged position at the center of the universe against all opposition, now there was no doubt that man was a mechanical microcosm

14 J. Donat, *The Freedom of Science* (New York: Wagner, 1914), pp. 43f.

who could only observe and record the smooth workings of the mechanical macrocosm. The hidebound Aristotelians had refused to look through Galileo's telescope claiming that anything they saw would only be caused by the distortion of a faulty lens. Now the mechanists threw out the secondary sense qualities by saying that they were only in the viewer and not in the object seen. Where Galileo's opponents had unjustly dismissed his arguments on a priori grounds, now the mechanists picked up the univocal wand and waved vitalism, teleology and critical realism out of the universe.

A final similarity is also striking. Just as the geocentric system finally toppled because of discoveries within astronomy and physics, so the mechanistic vision was cracked first from within physics itself.

The first compelling indication that the World Machine might have serious defects came in the last quarter of the nineteenth century. To summarize briefly, Faraday's study of electricity and magnetic force led him to conclude that magnetic action can be represented in terms of lines of force. Clark Maxwell, in 1873, worked out a set of differential equations that described the electromagnetic field and shortly thereafter Hertz devised several experiments which showed that Maxwell's equations represented reality: electromagnetic waves really exist. The electromagnetic theory was a clear challenge to the classical denial of action at a distance. Attempts were made to reconcile the field concept with mechanism; ether, the old and proven panacea, was invoked. But the fact was that the sort of ether which Maxwell's theory demanded proved stranger and stranger, less and less "material," the more physicists tried to imagine it.

The work of Faraday, Maxwell and Hertz was even more important in the sphere of epistemology. It hinted as what was soon to become clear: science does have limitations. The awareness dawned again that maybe reality was more than atoms in motion and that it was too rich to be expressed entirely and essentially in mathematical formulae.

Despite a number of problems raised by the new theory, Newtonians clung to their system. They pointed out that if there was such a thing as interaction between fields, it was taking place in Absolute Space and Time and according to Newtonian laws. And then came Albert Einstein.

We cannot here go into the interesting but involved contributions of Einstein to science. But we can at least point out the fact that his special and general theories of relativity accomplished a relativistic fusion of space and time with the result that the classical view of absolute space, motion and simultaneity fell from the category of scientific postulates. Moreover, by extending the concept of relativity to optical phenomena and maintaining the constancy of the velocity of light in a vacuum, Einstein showed that no *material body* can move as fast as light. He achieved a mathematical abstraction which required non-mechanical entities. He was able to geometrize about nature without saying that nature was nothing more than concretized geometry.

In addition, Einstein's adoption of Riemannian geometry to govern the curvature of his four dimensional continuum allowed him to work out a formula for gravitation differing slightly from Newton's law, which is based on Euclidean geometry. Newton's law satisfies the general requirements for accuracy but Einstein's formula accounts for several problems that Newton's does not, for example, the perihelion of Mercury, light from stars being deflected in the gravitational field near the sun, and the slower velocity of vibration of atoms on the sun when compared to terrestrial atoms.[15]

What was happening in science was the correction of classical physics, not its overthrow. The genius of Galileo and Newton was being furthered, not disregarded. The framework of classical science stood. It was the philosophy of scientism that was falling by the wayside.

The decisive inroad against mechanism was achieved by the

[15] See J. W. N. Sullivan, *The Limitations of Science* (New York: Mentor, 1949), pp. 60 ff.

quantum theory, originally devised by Max Planck in 1900, amplified by Einstein's theory of light quanta, and employed by Niels Bohr in the study of atoms. The main idea behind quantum is that energy radiates from a body not in a continuous flow, but in discontinuous packets or quanta. Bohr's research convinced him that the basic structure of matter depends upon the existence of quanta. Furthermore, contrary to the sanguine determinism of Laplace, he showed that it was seemingly impossible to predict individual atomic activities. Classical physics viewed atoms as tiny billiard balls that combined to form visible substances. Bohr followed Rutherford in conceiving the atom as a loosely-structured miniature planetary system in which electrons moved around a central nucleus. It made little difference whether the ultimate character of minute matter was wave or particle, the fact is that activity within the atom was inexplicable according to classical methods and laws.

In 1927, Werner Heisenberg established that the simultaneous knowledge of position and velocity necessary to realize the classical ideal was unattainable in microphysics. With Heisenberg's famous uncertainty principle, scientists came to the realization that

> Determinism is an idealization rather than a statement of fact, valid only under the assumption that unlimited accuracy is within our reach, an assumption which in view of the atomic structure of our measuring instruments is anything but realistic.[16]

Heisenberg's principle has very important consequences in the whole area of human knowledge. It destroys the artificial gap between knower and object known that had been an article of faith in the mechanistic system. The principle of uncertainty is based on the fact that in the very act of measuring sub-atomic particles,

---

[16] F. Waismann, "The Decline and Fall of Causality," *Turning Points in Physics* (New York: Harper Torchbooks, 1961), p. 113. For an explanation of why the uncertainty principle does not vitiate the concept of causality as understood in philosophy, see Alfred Stern, "Science and the Philosopher," *The New Scientist*, ed. P. Obler and H. Estrin (New York: Doubleday, Anchor, 1962), pp. 280 ff.

The measuring instruments of microphysics themselves exert a distinct influence upon that which they seek to measure, thus rendering its behavior in one or another respect unpredictable. The very attempt to observe a particle 'knocks it off its course' and the more accurately we pin down its position, for example, the more unsure we are of the degree to which we have affected its momentum. Among atomic physicists, in short, there are no innocent bystanders; the act of observation is at the same time unavoidably an act of participation.[17]

Human knowledge has again become important. No longer is science a completely objective, mechanical process that describes a world totally separated from man. The scientist as man enters into his science. He is a "participant-observer rather than a detached spectator."[18] The scientist must decide what particular aspect his experiment will reveal to him and he must be responsible for interpreting his findings as they are meaningful to him. He of course searches for scientific certitude, but he knows too that many of his investigations are strictly limited and can end only in one or another degree of probability.

Recognized limitations of method and the realization that "the creative scientist, whatever his field, is deeply involved emotionally and personally in his work," so that, "he himself is his own most essential tool,"[19] have brought about a new attitude in science. The spirit of post-modern science is well expressed in the principle of complementarity elaborated by Niels Bohr. As employed in quantum physics, complementarity refers to a duality of experiments in which the "wave character" and the "particle character" of sub-atomic entities are studied. Both are necessary for a complete explanation of matter, but they cannot be applied at the same time. They are mutually exclusive at any given moment, but complementary if applied successively. Likewise, the principle of complementarity is used in the study of position and velocity

[17] Floyd Matson, *op. cit.*, p. 143.
[18] *Ibid.*, p. 145.
[19] A. Roe, "The Psychology of the Scientist," *The New Scientist, ed. cit.*, p. 83.

of particles. Since they cannot both be measured at once, they are measured one after another and the results placed in a complementary scheme.

It is this principle that finally brings down the univocal attitude of classical physics. Mathematical physics is no longer seen as the monolithic idol of knowledge, the only gate to understanding reality. The many valid forms of human knowing and expression buried by the mechanists have been restored to their rightful share in the kingdom of intelligence. It is clear again that all of reality is open to complementary types of interpretation. The artist, philosopher, theologian, poet, mystic, and scientist, all contribute valid and meaningful insights to the common fund of human knowledge. This does not mean, of course, that science itself has accomplished the victory of free will over necessity or proved the immortality of the human soul. Scientific probes into the nature of matter are not to be confused with philosophical or theological issues.[20] But it does mean that science's abdication of univocal validity restores to philosophy and theology the right to stand on their own merits as valid approaches necessary for an integral understanding of reality.

With the downfall of mechanism, the whole scientific community is able to breathe a sigh of relief. The closed off world has been opened again. Mechanism has lost its hold on biology, psychology and ethics. Natural scientists no longer need to apologize to physicists for being unable to suppress their need for teleology in their study. The analytic, objective study of the human being does play a part in psychology, but no longer can it claim to be the whole answer. "The sum total of parts of a human being does not add up to a whole human being. Scientific objectification is painful to the patient and defeats our efforts to understand him." [21] Environment does affect human activity, but it does not

[20] See A. R. Hall and M. B. Hall, *A Brief History of Science* (New York: Mentor, 1964), pp. 308–10.

[21] Thomas Hora, "Existential Psychiatry and Group Psychotherapy", *Psychoanalysis and Existential Philosophy*, ed. H. M. Ruitenbeek (New York: Dutton, 1962), p. 141.

remove free will. Fortunately it has become unfashionable to challenge man's right and ability to achieve, by personal decision, his authenticity.

What does this all mean with regard to the relationship of philosophy and science? In answering this question, we must remember how philosophy suffered in the era of scientism. From Galileo to the twentieth century the nature and function of philosophy became more and more ambiguous. When the ability of the mind to attain objective and certain truth apart from mathematical physics was denied, philosophy was dismissed as a subjective clearing house for *a priori* assumptions. For centuries philosophers had been forced to take one of several courses: they could subordinate their thinking to science, openly oppose the philosophical assumptions of scientism in which case they would be little more than a voice crying in the wilderness, or they could retreat from the man-nature relationship into the comfort of introspection. It is hardly surprising that when, in the early years of this century, the need for a realist philosophy was realized again, philosophers were caught off guard. They had little or no experience at relating science to ultimate values. For all practical purposes, science and speculative philosophy were unrelated, completely divorced. This situation has had serious consequences. Science has prospered, no doubt. It has produced new drugs to prolong life and missiles to destroy life; it has achieved magnificent technical advances that have freed man from much subhuman labor. On the other hand, philosophy has suffered from a poverty of interest and new ideas. Its importance to mankind has been minimized. Because the findings of modern science have not been related to a suitable philosophy of nature, man has been denied a complete, integrated view of the universe in which he lives. In the pre-mechanical era the universe was romantic and meaningful precisely because the science of the time was related to contemporary moral and aesthetical values. The separation of science and philosophy has left unchallenged the Communist "philosophy of science" which claims that there is no God, that man is nothing more than a highly evolved form of matter, and that science is to be used primarily as a weapon of political revolution.

In recent years, with the spirit of complementarity, the situation has brightened considerably. Today we realize not only the necessity for but the possibility of achieving a sound philosophy of nature and a realist metaphysics. Philosophy itself is in the midst of a rejuvenation process thanks largely to the rise of existentialism and phenomenology. It is true that a great deal of thought must be given to determining the proper role of philosophy in the life of the mind and that fundamental disagreements remain among philosophers themselves as to how reality is to be approached and interpreted. But spurred on by the realization that they will be listened to, and that their task is of crucial importance to mankind today, the philosophical community shows definite signs of a new vitality that is most encouraging. The fact that there are different schools of philosophical interpretation is not necessarily a sign of weakness. For even within philosophy there is a complementarity of insights and mutual criticism that is beneficial to the entire philosophical quest.

Along with the revived interest in personalistic questions about man, his aspirations, perfections and goals, there is a growing awareness that "there is no science concerned with the whole of reality and knowledge. It is a specific field of philosophy. Any question which considers any phenomena scientific or other, in relation to the idea of the whole of reality and knowledge is a philosophical question." [22] Moreover, since the scientist is more than an observer of objects, since he is a participant in the very act of observation, his science is more meaningful to him on a personal level and involves him, willingly or not, in philosophical questions. Thus it is that an increasing number of scientists are professing their need for a philosophy that can serve as a framework for their investigations and a means of interpreting the ontological significance of their findings. The empirical data gath-

[22] Karl Stern, *op. cit.*, p. 286. See also Mortimer Adler, "Questions Science Cannot Answer," *The Logic of Science* (New York: St. John's Univ., 1964).

ered, the intuitions gained from experiential knowledge, the laws formulated on the basis of hypotheses: these must all be examined in the context of broader, more fundamental principles and values. Werner Heisenberg has written:

> The aim of research is no longer knowledge of atoms and their motion 'in themselves' separated from our experiential questioning; rather, right from the beginning, we stand in the center of the confrontation between nature and man, of which science, of course, is only a part. The familiar classification of the world into subject and object, inner and outer world, body and soul, somehow no longer applies and indeed leads to difficulties. In science, also, the object of research is no longer nature itself, but rather, nature exposed to man's questioning and to this extent, man here also meets himself.[23]

The next step, then, is to bring the scientist and philosopher closer together. How can this be done? If we are too practical, we cannot expect the scientist to go to the philosopher directly and study under him, or the philosopher to intrude upon the laboratory. Instead, a middle ground is developing called the philosophy of science and it is here that the philosopher and the scientist can and must become aware of each other's methodology, problems and answers. Men interested and knowledgeable in both disciplines are dedicating their labors to effecting a *rapprochement*.[24]

The time is right for a dialectic; the possibilities are limitless. Only a well-developed philosophy of science will enable the natural philosopher to go beyond the general principles of nature by giving him access to and explanation of the discoveries of science. Similarly, here the scientist will have an opportunity to evaluate the cognitive status of his science and to relate his knowledge to the ultimate, personalistic questions. I believe that the goal should be a single, unified philosophy of nature at once supported by

[23] W. Heisenberg, "The Representation of Nature in Contemporary Physics," *Symbolism in Religion and Literature*, ed. Rollo May (New York: Braziller, 1961), p. 227.

[24] One such group of scholars working for a *rapprochement* between science and philosophy is the Albertus Magnus Lyceum, located at the Aquinas Institute School of Philosophy in River Forest, Illinois.

and supporting the special sciences of nature and complementing the findings of mathematical-physics on the one hand and realist metaphysics on the other. Such an approach can give substance to philosophy and ultimate meaning to science. As Dubarle has written:

> In a sense, it is the privilege of modern science to oblige all mankind to ask itself as a body what is its being and destiny. Science no longer limits itself to providing food for individual reflection. Today the whole race of men is shaken by its revelations and finds itself faced with problems that cannot be avoided.[25]

Today there is no longer the wish to avoid these problems and, with the reconciliation of science and philosophy, there is a real possibility of answering them.

FAITH AND SCIENCE

A second issue of the Galileo case that is still a subject of controversy today involves the precise relationship between religious faith and modern science. Galileo was both a scientist and a believer; it was Galileo the scientist who wrote, Galileo the believer who recanted. But the lesson of his conflict with the Church is not that science and faith are essentially opposed. The lesson lies rather in its dramatic verification of what disaster can come to science or faith when either of these is extended beyond its proper boundaries and enters the domain of the other. A theologian *qua* theologian has no more authority in speaking about a matter of pure science than does a scientist in discussing Revelation and the Transcendent.

Today it is not so much a question of religion "threatening" science as it is of a lingering fear that science will somehow undermine the truths of religion, truths which we believe on faith. This change in emphasis has prevailed since the downfall of the Newtonian world-machine which seemed to tie the work of the Creator to a specific physical theory. Newton not only accepted the ultimate divine genesis and conservation of the universe, he also

[25] D. Dubarle, *Scientific Humanism and Christian Thought* (London: Blackfriars, 1962), p. 93.

assigned very definite duties to God in its daily running. As Professor Burtt has shown, it is possible to find many passages in Newton's writings which seem to teach that after the initial act of creation, God left the world to look after itself. But on closer investigation, as Burtt notes, we find that Newton had no intention of divorcing God from present control of and occasional interference in His vast cosmic machine.

Newton's religious beliefs led him to revolt against the absolute mechanism that would move God out of the universe. God must certainly be active in the universe and thus certain functions could be assigned to Him. Obviously, these functions would be those not otherwise explainable by Newtonian science. He put God on duty at all times "to prevent the fixed stars from collapsing together in the middle of space," and to keep the irregular motions of certain celestial bodies from creating chaos in the heavens. The important thing is that God was not only philosophically and theologically required to explain the universe; now he was *scientifically* necessary. God was the cosmic plumber who filled in the gaps and took over where Newton's laws left off. Leibniz saw the fallacy in this and remarked that he had little use for Newton's God—who apparently had done such a bad job of creation that He had to keep stepping in to repair it. The results of Newton's linking God to certain imperfections in the cosmos proved to be catastrophic for religion. It was not long before science was able to account with physical laws for these irregularities and Laplace could declare that there was no longer room for God in the universe. "I didn't," he said, "have to make use of that hypothesis." Tensions were bound to occur as scientists systematically removed God from the universe. "Defenders of religion" felt it necessary to do battle against the evils of "atheistic" science. The world of science and that of faith were completely separated.

With the advent of Darwinism and the application of mechanistic principles in biology, psychology and ethics, religious faith found itself robbed of its intellectual respect, crammed into a corner of the category either of anachronism or sentiment and left to die. Lord Kelvin said that only what he could see or feel was real. If that dictum is applied to God or grace, the result is bound to be the belittlement of belief.

On the other hand, there were many elements in the religious faith of the time that added to the difficulties raised by science's univocal attitude. Fundamentalism was still rampant. If the religious person had finally been able to get used to the idea that his earthly home is not at the center of the universe, he now had to face the supposed challenge of evolution to the creation account in the Book of Genesis. Many did so by denying the evidence of science with a cluster of Scripture quotations. Even more serious, perhaps, is the fact that many of the accidental aspects or untheological extensions of belief became static and essentialist. The common conception of God living "up there" somewhere, always ready to change His mind if enough prayers were sent His way, demanding adherence to the letter of the law, arranging everyday events to fit the desires of His faithful on earth: such an anthropocentric conception amounted to nothing less than a man-made God. Because superstition was often mixed with supernatural, when superstition was exposed, often what was truly supernatural was dragged down with it to be judged in the court of reason. To put it simply, much of the tension in the nineteenth century was due to a faith that had become crusted over with puritanical piety and superstitious practices. It was not just scientism that had to be overcome before faith and science could coexist again with equal rights; faith too needed to be purified.

Scientism has, on the whole, passed from the scene. The new science has succeeded in breaking down the myth that the doors of knowledge can be opened only with a mathematical passkey. Gone too is the once-popular image of the scientist as a man who, locked away in his laboratory or study, gathered facts but prevented his intellect from relating those facts to a personal philosophy; who, if he was a believer, prayed that his discoveries would not destroy his faith; who knew and cared nothing about art and culture and never experienced anything but precise, measured, intellectual satisfaction. It is highly doubtful that even at the apex of classical mechanism such a scientist ever existed. It is certain that none exists today. Even non-scientists understand that

> Science is not an occupation, such as working in a store or on an assembly line, that one may pursue or abandon at will. For a creative scientist, it is not a matter of free choice what he shall do. Indeed it is erroneous to think of him as advancing toward knowledge; it is, rather, knowledge which advances toward him, grasps him, and overwhelms him.[26]

The scientist loves truth wherever he finds it. The significant thing is that today he searches for it and finds it outside as well as inside his science.

Faith too has undergone a process of purification in this century. This is due largely to biblical criticism which has exposed the errors of attempting to make faith as rational as possible so as to conform to the "scientific" criterion of knowledge, or of considering faith as an act of the will only and therefore exempt from rational considerations.

When, at the end of the nineteenth century, historicism—which aimed at making history purely objective, demonstrative and scientifically certain—was applied to the New Testament, the results were disconcerting to say the least. Whatever did not meet the standards of the historicist was dismissed as the product of imagination and invention. Some Christian apologists attempted to defend the Scriptures on the challengers' home grounds. Adopting the basic assumptions of historicism, they eagerly set about trying to show that the Gospels are factual, chronological histories of the life of Christ. Their attempts reflect a false concept of what history is and, more important, what the Gospels are. The apologists could not answer the charge that the Jesus of history is not the Christ of faith because history as such and by itself is simply incapable of resolving the problem.

26 Gerald Holton, "Modern Science and the Intellectual Tradition," *The New Scientist, ed. cit.,* pp. 28f.

Next the nature and message of the Scriptures were examined with the method of Form Criticism. This method attempted to trace the origin and development of the pre-literary traditions in the Church which were eventually formulated in the Gospel writings. Following the lead of Martin Dibelius and Rudolf Bultmann, the form critics classified the materials contained in the Gospels according to forms or literary structures such as parables, prophetic or wisdom sayings, etc. Then, by studying rabbinical and parallel extra-biblical traditions as well as the *Sitz-im-Leben* in which the Christian tradition originated, they attempted to reconstruct and evaluate the Gospel sources. The method of Form Criticism is of limited value, but it has contributed to a better understanding of the Gospel tradition and therefore to a better understanding of the Gospels themselves.[27]

In addition, Bultmann's demythologizing principle has greatly influenced biblical studies. Bultmann accepts the Gospel presentation of Christ but insists that much of the New Testament worldview is out of date and needs to be demythologized, that is, interpreted in a manner meaningful to modern man who lives in a world revealed by science. Though one need not agree with many of Bultmann's conclusions, especially the gap that he leaves between the Jesus of history and the Christ of faith, he has succeeded in spotlighting the fact that the Gospel message and faith must be made relevant to our age.

These movements in biblical criticism have helped to clarify for us the objects, motives and the very nature of religious faith. For one thing, as Raymond E. Brown has pointed out,[28] we have been made to realize that the Gospels present primarily the historic (*geschichtlich*) Jesus, that is, Jesus as He had become meaningful to the believer with post-resurrectional insight. This is not to agree with Bultmann that it is almost impossible to

[27] See V. O'Keefe, "The Gospels Read as Gospels," in *Faith, Reason, and the Gospels*, ed. J. J. Heaney (Westminster: Newman, 1961), pp. 234ff.
[28] R. E. Brown, "Bultmann and the Post-Bultmannians," *The Bible Today* (November, 1964), p. 907.

discern in the Gospels the historical (*historisch*) Jesus as well.[29] What it does mean is that instead of viewing the Gospels as scientific histories recording the *ipsissima verba* of Christ, believers must see that "what we have in the Gospels is the witness of preachers and teachers who were not coldly objective reporters but ardently convinced believers, and whose interest was salvific and not biographic."[30]

In the Gospels, facts and doctrine are linked together; they are testimonial documents, the witness of men who lived with Christ and to whom He revealed Himself. There is a valid and necessary apologetic task, for the Christian message has a certain natural credibility.[31] But it must be remembered that the Gospels were written *ex fide ad fidem*; they call for a response of faith and men are offered faith, not argued into it. The return in Protestant and Catholic theology to the biblical presentation of God and man in salvation-history has helped us understand better who God is by giving us a clearer knowledge of what He is not. God is the Personal Creator who sent His only Son, Jesus Christ into the world to die and overcome death that all men might be saved through personal commitment to Christ. He is not a celestial mechanic, or a disinterested moral scorekeeper. We are aware again that faith is a personal response not to scientific demonstration nor to irrational feeling but to a loving God who has revealed Himself to us and given us the grace to accept that revelation. Both knowledge and love are essential to faith because both are essential activities of the human person, and in faith it is the human person who gives himself to his God.

[29] "The adventures of non-Catholic biblical criticism over the past century make it evident that he who rejects the Christ of faith will soon end up by reducing the Jesus of history to a pale figure without religious significance. Conversely, he who makes light of the flesh-and-blood Jesus of history in the name of a more spiritual faith will end up prostrating himself before a timeless myth. If we are to be true to the Gospels we shall insist on retaining both fact and interpretation, both history and faith." A. Dulles, *Apologetics and the Biblical Christ* (Westminster: Newman, 1963), p. 41.

[30] R. E. Brown, *loc. cit.*

[31] For an excellent contemporary apologetic work, see A. Dulles, *Apologetics and the Biblical Christ.*

After centuries of distortion by the magnetic tug of rationalism on the one hand and fideism on the other, we have returned to the traditional notion of faith as conceived by the Church Fathers and the great medieval theologians. It will repay us here to analyze more closely the nature of religious faith.

In general, to have faith means to assent to the truth of something on the word of another. It is not unreasonable to accept the testimony of trustworthy witnesses; we do this many times a day, for example, in reading a newspaper, listening to a teacher, talking to friends, etc. To have religious faith is to assent to certain religious truths because they are revealed and guaranteed by God Himself. By faith one holds as true what God has revealed about Himself not because he sees its intrinsic truth with his human mind, but because of the authority of God. Faith leads man to make the intellectual judgment that what God has told man is true, for example, that He is three Persons in one God. A person judges that this is true not because he can demonstrate it by his reasoning powers, but because, moved by God's grace, his will directs his intellect to accept the truth of God's revelation.

Because the truths which a Christian believes on faith are essentially supernatural (for example, the Trinity, the Incarnation and the Redemption), they cannot be arrived at by human reason apart from faith, nor can they be understood fully by human reason even with faith. They are mysteries far too profound for the human mind to understand completely in this life. But these truths, even though comprehended only imperfectly, give us a deeper understanding of the life of God and the relationship of man to God than any we might have arrived at by the power of human reason alone. If belief is the only way for man to come into contact with supernatural mysteries, it is not unreasonable for man to believe. St. Thomas Aquinas remarks:

> Someone may say, 'Is it not stupid to believe what is not seen?' In answering this, I say first of all that the imperfection of our understanding takes away the force of this difficulty. As a matter of fact, if man could know perfectly all things visible and invisi-

ble, it would be stupid to believe what we do not see. However, our knowledge is so imperfect that no philosopher has ever been able to make a perfect investigation of the nature of one fly. We read that a certain philosopher spent thirty years in solitude, so that he might study the nature of a bee. If our intellect is so feeble, then, is it not stupid to refuse to believe anything about God other than what man can know by himself? And so, against this objection, it is stated in Job 36:26, 'Behold God is great, exceeding our knowledge.'

In the second place, it can be replied that if some master teacher said something within the area of his own science, and if some unlearned person said that the master's teaching was not so because he did not understand it, one might grant that this person would be considered rather stupid. Now it is obvious that an angel's understanding surpasses the intellect of the greatest philosopher far more than does the intellect of the great philosopher surpass that of the unlearned man. So the philosopher is unwise if he refuses to believe what the angels say and much more so if he refuses credence to what God says. Again, against this objection stands the statement of Sirach 3:25. 'For many things are shown to thee above the understanding of men.'

Thirdly, it can be answered that if a man refused to believe anything unless he knew it himself, then it would be quite impossible to live in this world. How could a person live if he did not believe someone? How could he even accept the fact that a certain man is his father?

A man must believe something regarding things he cannot know by himself. But no one is as worthy of belief as God. Those who do not believe the truths of the faith are not wise; rather they are foolish and proud: as the Apostle says (I Tim. 6:4), 'He is proud, knowing nothing.' For this reason he also says (II Tim. 1:12), 'I know whom I have believed and I am certain,' and Sirach 2:8 reads: 'You that fear the Lord, believe him.'

In the fourth place, one may also reply that God proves that the teachings of the faith are true. Suppose a king sent a letter stamped with his own seal. No one would dare deny that the letter was sent with the king's approval. Now it is clear that all the things that the saints believed concerning the faith of Christ, which they have handed down to us, are marked with God's seal. This seal is manifested by those works which no mere creature can perform. These are the miracles whereby Christ confirmed the statements of the apostles and saints.[32]

[32] St. Thomas Aquinas, *Sermon on the Creed*, trans. L. Shapcote in *The Three Greatest Prayers* (London: Burns, Oates, 1937), pp. 40f.

The act of faith is above reason, but not against it. Reason can show the credibility of the fact that God has spoken to man. It can, after thorough investigation, arrive at the intellecual conviction that God's revelation in Sacred Scripture and Sacred Tradition is free from contradiction, that it is in the conformity with right reason, and that it is supported by an impressive list of miracles and fulfilled prophecies. Above all, reason can encounter the presence of God in history, as He now manifests Himself in the Christian community, which is the witness of Christ, and which, by its unity in charity, faith and hope, its fidelity to God, and in its living continuity through time and space, manifests God's work in the world. Even in the failures, sometimes disastrous, that manifest its human frailty, reason can see that this community, as human as any other, could not transcend its own limitations without divine support. But human reason can go no further than to recognize in this witness the call of God. It cannot accept as unquestionably true the teachings of revelation unless the human will, acting freely under the grace of God moves the intellect to assent firmly and irrevocably to these suprarational truths. Faith is a gift of God.

The act of supernatural faith is free. The will, moved by the grace of the Holy Spirit, in turn moves the intellect to assent. It is a free and personal act; it cannot be coerced. Yet, because faith is a gift, man cannot without God's help commit himself to supernatural truth. Here one can speak of faith as a "leap into the dark," in the sense that reason necessarily halts at the threshold of the supernatural, certain of how it got there, but needing elevation to go on to the final commitment.[33]

The notion of faith as we have described it has important corollaries. First of all, the role of reason is presupposed to faith. Reason can give serious arguments for the existence of God; it can establish the credibility of the fact of a divine revelation; it can provide the concepts that will be used analogically in formu-

[33] See M. C. D'Arcy, *Belief and Reason* (London: Burns, Oates, 1946), pp. 37 ff.

lating the Christian mysteries. But only if one is willing to accept the gift of faith which God offers, will he be able to believe and to make his own God's revealed truth.

Secondly, faith is not a static, sterile assent to God as He has revealed Himself to man. It is a dynamic commitment, a personal involvement. The Christian message remains unchanged, but the believer does not. True faith is evidenced in action. Faith leads us to love God, the object of our faith, and to witness to that love in our lives. God is our "ultimate concern," He is our ultimate practical truth. Faith is the decision to accept that personal truth and to order our whole being and action to Him who is truth itself.

If one accepts even the broad outlines of this notion of faith, the solution of the problem of faith and science will begin to fall into place. Though faith and science lead to very different kinds of knowledge, they are not opposed, nor is either less real than the other. The physicist can discover the laws that describe motion, the biologist is able to discern and relate factors involved in the growth of living things, the psychologist may describe the place and content of preconscious, subconscious and conscious forms in the process of human knowledge. A natural philosopher is able to demonstrate, arguing from principles of reason, why this natural agent acts as it does and the metaphysician can even tell us what it means to exist and to have free will. An artist may capture beauty and truth on canvas and a mystic can experience the inexpressible. The man of faith looks at reality from above. Faith does not close one off from the universe but opens up "a vast universe of another kind, which leaves intact for man but transfigures with new meaning the universe before his eyes." [34] Faith and science are complementary approaches to reality. It is well to remember when tensions occur that

> ... Faith cannot oppose reason; it can only oppose other *faiths* which are being used as rival conditions or presuppositions of

[34] Charles Davis, "Faith and Reason," *Faith, Reason, and the Gospels, ed. cit.*, p. 9.

reason such as the faith-principle of Marxism, or scientific humanism, or anti-Christian rationalism.[35]

Of course faith and science are autonomous; each has its own principles, method and formal object. But this autonomy should be the starting point for a dialectic and not a defensive wall. The scientist who is a Christian knows this well. He is not shocked by the immensity of the universe, his faith tells him that God is infinite and transcendent. The fact of man's material insignificance in the cosmos does not paralyze the believer, he knows well his own nothingness before God and the gratuity of His saving love. The order in nature and yet the basic mystery in matter do not leave him dumbfounded; he believes in the divine creation and conservation of all being in the universe. For the scientist-believer,

> The study of nature is a journey towards the divine ideas of the Creator, a quest for natural revelation: a worship at a new altar, but dedicated to the same God.[36]

Because a scientist's faith is a part of his person, to the extent that he himself enters into his work, his faith is relevant to science. In fact, they feed each other in his meta-rational thought processes. The more science he knows, the deeper can be his experience of faith. His faith can supply him with certain categories of thought in which his scientific and philosophical insights are enriched without being confused, in which they are related to the ultimate Source and Goal of all being, the personal God to whom he is personally committed.

Science and faith have both come a long way since the time of Galileo. Today's scientists do not reject technology, nor do they fail to use analysis. But they recognize the necessity of unifying knowledge. After breaking things up through analytic technique, they insist on putting them back together and this involves them in intellectual complementarity. Even a non-believing scientist knows that he cannot rule out religious faith on the basis of his

[35] Alan Richardson, "Faith and Presuppositions," *Faith, Reason and the Gospels*, p. 83.

[36] Friedrich Dessauer, "Galileo and Newton," in *Spirit and Nature* (New York: Pantheon, 1954), p. 319.

science and that if he denies religious faith, he does so gratuitously. Theologians, having survived the excesses and culled the advances of analytic biblical criticism, have been impelled to restore faith to its integrity and its vitality and to apply the Gospel message to the personal needs and possibilities of twentieth-century man.

We must not imply that no difficulties remain. Scientists and religious thinkers are still, and will always have to be, cautious in their respective approaches to each other. But there is a growing awareness that each is seeking and finding truth and this provides the foundation for mutual respect. Many scientists realize that

> Science cannot contribute significantly to the renewal of mankind unless it is practiced with ultimate concern. We realize too that the term 'ultimate concern' may be translated into 'love of God and love of neighbor.' But it is not clear what this means with respect to our day-by-day work and thought, and how this can enter into the rationale of our methodology.[37]

And religious spokesmen are willing to admit that

> In many cases there exists between science and religious truth a barrier of misunderstanding for which religious thought is not entirely without blame, and which it alone, in any case, can overthrow. The mind in love with God must also be able to assume into the process of its own growth all the valid acquisitions of human knowledge. At the present moment, we cannot say that it does this with complete success. There is work to be done. It requires an effort which will often be hard to maintain and which will call for delicate adjustments.[38]

Undoubtedly there will always be some pockets of tension in the dialectic between science and faith. But man in search of salvation and attempting to achieve his true dignity can no longer afford the schizophrenia of having one world of faith and another, completely different world of science.

[37] Harold K. Schilling, "A Contemporary Macedonian Plea," *The Union Seminary Quarterly Review*, v. 28, no. 2 (Jan. 1963), p. 121. Dr. Schilling is Dean of the Graduate School and Professor of physics at Pennsylvania State University.
[38] D. Dubarle, *op. cit.*, p. 119.

The time of isolationism is past, both for the spirit and for religion, for, through the body, we are bound to the cosmos as a whole. Only in union with it is the human spirit meant to be perfected and to find its shelter in the sphere of the divine.[39]

Science and religion must work together with mutual respect and assistance. In the following chapter, I hope to show at least the possible direction such a cooperative effort might take.

[39] Hans Urs Von Balthasar, *Science, Religion, and Christianity* (London: Burns, Oates, 1958), p. 48.

CHAPTER VIII

The Church and Science

When science broke away from the abuses of theological domination, the Catholic Church, which had for so long fostered the advance of reason as an essential factor of human perfection and the key faculty in the service of faith, now found itself on the defensive against the dominative tendency of human reason. The vision of Augustine, the monastic preservation of knowledge during the early Middle Ages, the massive synthesis of Aquinas and labors of Christian humanists such as Erasmus all seemed nullified in the divided worlds of science and faith.

Through the end of the nineteenth and into the early years of the twentieth century, many Catholic theologians were engaged in writing polemics against those who attempted to replace or dilute religion with the various species of materialism and agnosticism that spontaneously generated from scientific optimism. In general the Church's attitude toward science was cautious and apologetic. And though the trend is in the right direction today, reconciliation if still not a reality.

There are still some who applaud the divorce of science from the Church. But there is real tragedy here. Intellectual schizophrenia is never a happy state of affairs. By failing to appreciate the proper role of science in the overall picture of man and the universe and to develop a positive theology of the temporal order, theologians were unable to relate the findings of the scientific community to the cosmic action of God. As a result, believing scientists felt a tension between the knowledge and love of God as spoken of by theologians and as they experienced Him in their scientific work.

It was left to a man who was both a priest and a scientist to bring this tension to light and to offer his life and work as a sacrifice on the altar of resolution. There is a sense in which Father Pierre Teilhard de Chardin was a prophet. He called on scientists and theologians to repent of their proud independence and saw their cooperation as an essential feature of continuing evolution in the Noosphere. He blazed a new trail, experienced a new vision, and tried to forge a new language to communicate it.

The passion of Father Teilhard's life was to unite his scientific knowledge and love of material creation with his faith-knowledge and love of God. To an extent he succeeded. The impact of his writings indicates that his personal synthesis has a meaning for all men. He left us the outline of a combined scientific-philosophical-theological view of man and the universe. More important, perhaps, he gave us a vision, at once cogent and compelling, of mankind evolving under grace slowly, steadily toward the Omega-Point where he would know, love and accept nature, know love and accept himself, know love and accept God, the Alpha and Omega, the source of all being.[1]

Teilhard's system has aspects that appeal to scientists, philosophers, and theologians, and some that do not. The creative scientist can appreciate Teilhard's personal involvement in his own scientific endeavors, his seeking to understand the universe not by analysis only but by union or connaturality with the universe to

_____
[1] See Bernard Towers, "Jung and Teilhard," *Teilhard de Chardin: Pilgrim of the Future*, ed. N. Braybrooke (New York: Seabury Press, 1964), p. 88.

which he belonged. The scientist can understand how Teilhard's science introduced him to and was enriched by the aesthetic continuum in which he recognized that he could not be what he was without the presence of all he encountered nor could all he encountered reveal itself fully unless he not only studied it, but identified with it and loved it. Great scientists, as Loren Eiseley has said, combine an aesthetic and religious response to nature with a passionate inquisitiveness of research.[2]

A philosopher can see in Teilhard's works a contemporary reconsideration of the perennial problems of philosophy: Matter and Spirit, Body and Soul, Person and Society, Cosmos and God; moreover, Teilhard's understanding of the principle of finality is in the same line as and perhaps more developed than that of Aristotle and St. Thomas Aquinas.[3] But professional philosophers have complained that his methodology is not clearly defined, that he mixes at times philosophical, scientific, and theological data, and that his whole line of argument lacks a metaphysical structure.

Catholic theologians are likely to be pleased with Teilhard's vital integration of faith and science and his fine balance of incarnational and eschatalogical emphases. He truly helped open the cosmos to theological penetration. But many object, he did not bring into his systems with sufficient attention the doctrines of creation, the fall, and redemption; and the place of grace and free will in the process of evolution are not clearly established.

Whatever attitude one adopts toward Teilhard, skeptic, critical, or enthusiastic, one thing is certain: he has raised questions that stand at the core of the spiritual destiny of mankind. It is only fair to say that he did not intend to elaborate a complete and final synthesis of science and theology; he knew that his theory would need expansion, precising, and correction. What he wanted

[2] See C. P. Snow, "The Moral Un-Neutrality of Science," *The New Scientist*, ed. P. Obler and H. Estrin (Garden City: Doubleday, 1962), pp. 127 ff.
[3] See J. L. Russell, "The Principle of Finality in the Philosophy of Aristotle and Teilhard de Chardin," *Heythrop Journal*, vol. III (1962), pp. 347 ff.

to communicate was a vision based upon his passionate desire to relate the truths of science, philosophy and faith. Teilhard was aware that his views did not fit easily into accepted hierarchical structures and he had the humility to accept the fact that they would have to be introduced gradually. He knew that his religious superiors might have reason to prohibit, temporarily at least, the publication of his writings. He wrote to a friend:

> There remains the fact, and I acknowledge it fully, that Rome can have its reasons for considering that, in its actual form, my vision of Christianity is premature or incomplete and that, as a consequence, it cannot be diffused at present without objections.[4]

To those who suggested that he leave the Jesuits so that he could have freedom of publication, Teilhard replied:

> I would believe myself guilty of betraying the 'world' if I withdrew from the position assigned to me in it ... The Society is not less but more and more precisely my point of insertion and of work in the universe.[5]

Yet Teilhard was convinced that his vision belonged to the world and that it needed to be submitted to the judgment of his peers. It was in accord with his wish that a committee of world-renowned scholars worked together to oversee the publication of his writings after his death.

It is a tribute to Teilhard that today there is a fresh and urgent call for Christian humanism and cosmic Christianity. Theologians and scientists alike are under a Chardinian imperative to put away their stereotyped versions of each other's vocation and to come to the table of dialogue with open minds and a sense of solidarity. Scientists are likely to be as surprised to learn about the current trends in theology as theologians will be at the mature attitude of the new science.

The entire science-theology relationship, it seems to me, is in a

---

[4] Teilhard de Chardin, letter to the Superior General of the Society of Jesus, dated October 12, 1951, reprinted in *Teilhard de Chardin: Pilgrim of the Future*, ed. cit., p. 106.

[5] Cited by C. E. Raven, "Orthodoxy and Science," *Teilhard de Chardin: Pilgrim of the Future*, ed. cit., p. 58.

new context. Difficulties that have kept the two disciplines divided no longer allow of oversimplified answers.

Theologians can no longer justify a hostile attitude toward science because it seems to lead men away from religion by reducing the need for God as an expanation, or because, by seeming to limit the area of the explicitly sacred, it appears to destroy the immanence of God in the universe.

The new science admits the possibility (some would even say the necessity) of an ultimate ordering principle that might even be called a "grace-principle." Teilhard suggests that we have moved from a religion of science through the crisis created in physics by the discovery of radioactivity, in biology of the refusal of life to reduce to mechanical components, in sociology of the increasing obscurity of human destiny, to a time when we need a new mystique of religion in science. The scientific idea of an evolving world, he says, creates a milieu favorable to a better understanding of the Incarnation. An ascending anthropogenesis helps us understand the descending rays of Christogenesis. Though science cannot itself prove or disprove the existence and action of God, it seems to support rather than reduce the concept of divine immanence and activity in the cosmos.[6]

Many theologians have been less than enthusiastic about the ultimate value of science because they conceive of science as an analytic, mechanical approach to nature which completely abstracts from the impact of technological advance on the moral life of man. Besides being wrong in their conception of what science is and does, they imply that technology has led man to a frantic pursuit of comfort and fostered a lack of personal involvement in society. But if this is true, one would have to show that man in society has fallen from a more serious commitment, one that would be obvious in the pre-technic age. Certainly men

[6] See *The Evidence of God in an Expanding Universe*, ed. J. C. Monsma (New York: Putnam, 1958); *Science and Religion* by the same editor (New York: Putnam, 1962); E. L. Mascall, *Christian Theology and Natural Science* (New York: Ronald Press, 1957); and P. Chauchard, *Science and Religion* (New York: Hawthorn, 1962).

have misused technology, but that is not the fault of technology. If it is true that scientific progress has provided mass resources for a psychic life lived on the level of boredom and a spiritual life lived on the level of Kierkegaard's aesthetic man, what is needed is not less scientific progress, but a corresponding advance in the teaching and living of Christian ascetic theology. Leisure time and material advances will be ultimately meaningful to man if it is made clear how contemplation, prayer and love can best be fostered in contemporary society. It is not a matter of renunciation only, but of positive adoption of the advances afforded by technology to further the life of God, man and the universe. Theologians can no longer be comfortable with a straw-dummy concept of science and the scientist. The very concern of scientists with the implications of their discoveries is an appeal for theological assistance. The *aggiornamento* begun by Pope John XXIII carries with it the command to make Catholicism relevant and vital to man in this scientific era.

A third idea that must be seen in its new context is that of science as fostering among mankind a Promethean self-image. There is a danger that the conquests of science could lead to naturalism and Utopianism. The human race could conceivably fall into a new sin of hubris. But the spirit of the new science is not Laplacean. Scientists are profoundly aware of the mystery of nature which they seek to penetrate; they reject the scientism of a bygone age and are open to and asking for complementary views of the cosmos. The very possibility of a corporate sin of pride should stimulate theologians to even greater efforts in their vocation as witnesses to the Good News of salvation. Theirs is the duty to remind the world of its dependence upon God and to preach to the human community not only "Without Christ we can do nothing," but also, "We can do all things in Him who strengthens us."

If Catholic theologians have had a somewhat hazy idea of the nature of contemporary science, it may also be true that scientists have not been exposed to the riches of Catholic theology and its relevance to them precisely as scientists. By bringing out the cosmic significance of four central truths of the faith, perhaps we can indicate something of a specifically theological understanding of the cosmos and the scientist's work in it.

CREATION

The Old Testament portrays creation as the first of many acts in the history of salvation. Creation is a free act of God and everything created shares partially in His perfection. Matter is like the clothing of God's thoughts, for all created realities, living or inanimate, express a creative idea. God's one act of creation is willed from all eternity. But, in time, it is not a past and fully accomplished event. God transcends time but is personally involved in each moment of time. He is intimately present to all creatures conserving their being and directing their activity.

> To say that the world is created is to say that it does not contain within itself the cause of its existence; that it does not securely possess its existence in a complete and permanent way; that its motion is a breathless and ceaseless striving toward the Being who keeps it in existence. At every moment the world receives its existence from the divine intellect and will. Every moment it drops into the past and arises from the future within the present. The moment is really the heartbeat of created things. It is a sign of their plenitude, since in it the Creator brings the world into existence. It is a sign of their finiteness, frailty and transience, because in it existing things must be renewed and maintained in existence ... each moment is a visible sign of the continual genesis of the universe.[7]

The universe as creature is sacramental; it speaks to us of God's presence. Not only in its being, but in its activity, creation praises God. The very inclination of nature toward its own perfection is planted there by God and is ultimately an immense aspiration to God, expressing a natural love of God.

Evolutionary theory conceives of the universe almost as a creature constantly struggling toward more perfect organization,

---

[7] Jean Mouroux, *The Mystery of Time*, trans. J. Drury (New York: Desclee, 1964), pp. 41f.

toward a more developed nervous system, a more perfect brain, toward consciousness, toward man. Scripture too represents man at the height of creation. In Genesis man is appointed God's vicar over creation. He is given the sacred duty to "fill the earth and subdue it." He is linked with nature. but to rule, complete, and achieve it. He commands nature in the name and the service of God. In Mouroux's phrase, man "is truly creation's priest. And fraternal nature, not unhelpful, but seeking, desiring, looks up to him alone who can fulfill her desire by giving her a soul and a voice with which to honor her God." [8]

So intimate is man's brotherhood with nature that when Adam and Eve sinned and the relationship of man to God was disrupted, the perfect harmony of material creation was also lost. In a sense, the sin of man redounded to the core of creation and the perfect subjection of nature and beast to their human mediator turned to defiance in imitation of and following upon his sinful rejection of God. After the primal sin, material creation still reflects the glory of God but now it does so mutely and with difficulty; its priest, the voice through which it could give vocal praise to God, has been unfaithful to his priesthood. With sin, suffering and death enter the world and war is declared between the spirit and the flesh. Nature, holy in itself, is now pliable to evil ends and able to serve sin as well as sanctity. "The soul, the body, and the world—how terrible a continuity, how wide a field for the ravages of sin!" [9]

THE REDEMPTIVE INCARNATION

God did not abandon creation in its misery. In the fullness of time, He sent His only-begotten Son, Jesus Christ, who united a human body and soul to his divine Person and entered the universe to bring renewal, healing, and, in fact, a new creation. The Church teaches that while creation is properly an act of the Holy

[8] Jean Mouroux, *The Meaning of Man*, trans. A. Downes (Garden City: Doubleday, 1961), p. 34.
[9] *Ibid.*, p. 36.

Trinity, it is attributed to the Father who created through His Word. In the Incarnation, it is this Word of God who became man. The creative Word who was in the beginning with God now entered into space and time bringing a new creation.

Through the humanity of Christ, God's love was brought to man in visible form. "The man Jesus as the personal, visible realization of the grace of redemption is the sacrament, the primordial sacrament, because this man, the Son of God Himself, is intended by God the Father to be in His humanity the only way to the actuality of redemption." [10] Matter was given a share in the redemption itself. Christ's assumption of human flesh, his use of bread and wine in the first eucharistic banquet, and the references to material creation in His sermons bringing transcendent and spiritual realities to the minds and hearts of men, gave to matter, over and above its dignity as a reflection of the Creator, a participation in the very accomplishment of redemption. Christ's resurrection from the dead was the sign of His victory over sin: the triumph of the new creation. And it was the pledge that matter will share in the final gift of eternal glory.

Because Christ is God-made-man, he has assumed the headship of all men; He is King of the universe. This means that Christ Himself as man must share, and preeminently, man's role as mediator between Creator and creation. Far from abrogating the command given to mankind to subdue the earth, the Incarnation gives that command new meaning and the possibility of fulfillment. There is good reason for Teilhard to envision all of nature moving toward Christ the Omega-Point, for it is through Christ's humanity that creation will finally be returned to the Father.

The redemption has been accomplished, but the conquest needs to be completed and the new order brought to perfect manifestation. Creation is still ambiguous; it is subject to man and it reflects his struggle toward sanctity. When he advances, it advances; when he falters, it is at the service of sin. The grace of Christ that

[10] E. J. Schillebeeckx, *Christ the Sacrament of Encounter with God* (New York: Sheed and Ward, 1963), p. 15.

transforms man allows man to transform the universe. As man's love of God, his neighbor and the cosmos becomes progressively purified, nature will be more and more "an instrument of praise and benediction in the hands of the sons of God." [11] Thus Teilhard believed that the law of complexity-consciousness now operates on the level of spirit and that this law will be transformed by man's freedom into a growth of love centering more intensely on the ultimate personal pole of convergence, the Omega-Point. It is almost as though the cosmos, under grace, continually pressures man to live up to his priesthood of creation by freely identifying himself with the movement toward the source of love, Christ Himself.

THE SACRAMENTAL NATURE OF THE CHURCH

In contemporary Catholic theology there has been a re-emphasis on the fact that the Church, the Mystical Body of Christ, is the continuation of the bodily presence of the heavenly Christ in the world. The very nature of the Church as a visible sign of redemption means that her principal acts, the sacraments, bring Christ's eternal saving action to the world.[12] Christ in His Church makes Himself visible on the earth, ready to encounter man with his redeeming personality, an encounter initiated by Christ in the sacraments.

The central event in the Church is the celebration of the sacrament of the Eucharist, the re-presentation of Christ's redemptive death on the Cross. Man brings to the altar of sacrifice bread and wine, made by human effort from the fruits of nature. These are taken by the priest who, empowered by Christ in the Church, utters for Christ the words of consecration so that the bread and wine become really, substantially, and truly the Body and Blood of Jesus Christ—the living Christ who is at the right hand of the

---

[11] Mouroux, *The Meaning of Man, ed. cit.,* p. 37.

[12] The activity of Christ in the world is not restricted to the sacraments, since Christ can confer his grace as he sees fit. But he instituted the sacraments as visible manifestations of his grace and the ordinary means of incorporation into the full life of his Mystical Body.

Father. The Church joins with Christ in offering to the Father the perfect act of worship: his redemptive sacrifice accomplished once for all time on the Cross.

Father Edward Schillebeeckx brings out the relation of Christ to the world by virtue of the Incarnation, the Church and the Eucharist:

> A close unity exists between "inward" and "outward" grace, but the whole created world becomes, through Christ's Incarnation and the God-man relationship which is consequent upon it, an outward grace, an offer of grace in sacramental form.[13]

The Church as a continuation of Christ's earthly existence presents a special presence of Christ in the world:

> ... the preaching and the sacraments of the Church can be regarded simply as the burning focal points within the entire concentration of this visible presence of grace which is the Church, for thanks to the Eucharist, Christ is really *somatikos*—physically —present in her, and because of this physical presence, also personally present.[14]

The Eucharist, of course, was of singular importance in Teilhard's thought. For Teilhard, it is first by the Incarnation and next by the Eucharist that Christ organizes us for Himself. By His Incarnation, He inserted Himself not just into humanity but into the universe that supports humanity, and not simply as another connected element, but as a directing principle, a center toward which everything converges in harmony and love. In the Eucharist, Christ is "a universal Element" and through His eucharistic presence, He controls the whole movement of the universe. Referring to this presence, Teilhard writes:

> Its energy necessarily extends, owing to the effects of continuity, into the less luminous regions that sustain us ... through our humanity assimilating the material world and the Host assimilating

---

[13] E. J. Schillebeeckx, *op . cit.*, p. 21.

[14] *Ibid.* "Physically" here does not mean that Christ's presence in the Host is able to be detected by the senses or that he is present in a quantitative manner. Rather, he is present sacramentally in his Sacred Humanity in a substantial and personal manner.

our humanity, the Eucharist transformation goes beyond and completes the transubstantiation of the bread on the altar . . . In a secondary and generalized sense, but a true sense, the sacramental species are formed by the totality of the world and the duration of creation is needed for its consecration. 'In Christ we live, and move and have our being.' [15]

The influence of Christ is not artificial or arbitrary, not purely juridical or moral, but is truly at work in the reality and immensity of the cosmos.

Teilhard did not always draw an explicit and exact distinction between the natural and supernatural orders, between nature and grace. But he was keenly aware that some would think that he was positing a supernatural term for the natural process of evolution. Thus he wrote:

Christ is, of course, not the center which all things here below could naturally aim at embracing. Being destined for Christ is a favor of the Creator, unexpected and gratuitous. It nonetheless remains true that the Incarnation has so recast the universe in the supernatural that, concretely speaking, we are no longer able to seek or imagine the center toward which the elements of the world gravitate without the elevation of grace.[16]

There is strong theological support for this view. As Karl Rahner remarks, "one has only to ask why a supernatural end can be set for man without annulling his nature, and why God cannot do this with the nature of something below man." [17] The axiom of Aquinas that grace does not destroy, but presupposes nature and perfects it means that the whole reality of natural man: intellect, free will and activity, is brought to a new perfection by the gift of grace and led to a goal it could never of itself attain. By accepting evolution and its consequences regarding the relation of man to the universe, Teilhard has extended this axiom to the whole cosmos. Christ's presence in the Church

[15] Teilhard de Chardin, *The Divine Milieu*, trans. B. Wall (New York: Harper and Row, 1960), pp. 114f.
[16] Teilhard de Chardin, *L'Union Créatrice*, p. 14, cited by Christopher Mooney, "Body of Christ in the Writings of Teilhard de Chardin," *Theological Studies*, vol. XXV (December, 1964), p. 583.
[17] Karl Rahner, *Theological Investigations* (Baltimore: Helicon, 1961), vol. 1, p. 317.

and especially in the sacrament of the Eucharist, is the active Center radiating spiritual energies that lead the universe back to God.

THE PAROUSIA

If science and philosophy can supply Teilhard with a vision of mankind at the head of the universe moving in a communion of desire and under the impulse of spiritual energy toward an ultimate personal term, theology can verify that this ultimate Center of convergence must be Christ, for it is Christ who will bring all to perfection in the Parousia, His triumphant return to the earth on the last day. Christians are able to anticipate the end of the world because they see it not as a catastrophe, but as the final return of all things to God through Christ. All of creation moves toward this climax and liberation. "For the creation waits with eager longing for the revealing of the sons of God" (Rom. 8:19).

Exactly what will take place at the Second Coming of the Lord is a mystery. There are many texts in the New Testament that refer to the Parousia but they are generally vague and subject to a variety of interpretations. It is certain that Christ will judge the living and the dead on that day and "they who have done good shall come forth unto resurrection of life, but those who have done evil unto resurrection of judgment" (Jn. 5:29). We also know that "the promised restoration which we are awaiting has already begun in Christ, is carried forward in the mission of the Holy Spirit and through Him continues in the Church in which we learn the meaning of our terrestrial life through our faith, while we perform with hope in the future, work committed to us in this world by the Father, and thus work out our salvation." [18]

Relative to the cosmos, one may speculate that the end of the world will not be its destruction, but its transformation; the universe will not be annihilated, but made new. "Behold I make all

[18] *Constitution on the Church* by the Second Vatican Council, November 21, 1964, Chapter VII, N.C.W.C. edition, p. 55.

things new." (Apoc. 21:5). The power and splendor of Christ will penetrate all of creation and this leads Thomas Merton to say that "Christ will not only appear on the clouds of heaven in judgment but will also, at the same time, shine forth through the transfigured trees and mountains and seas of a world divinized through its participation in the work of His kingdom." [19]

In an incisive passage, Alois Winklhofer relates Christ, creation, and the Parousia.

> On this day of the Lord, His inherent power of inducing a new mode of being is given full scope, for He is, indeed, the Lord of the world, its principal element, the ferment most active within it. Everything receives His seal and becomes 'conformable to Him' (Rom. 8:19). The transformation we speak of is not simply ethical and spiritual, but reaches to the very essence of all creation, which, in consequence, becomes a clear and luminous revelation of the 'image of the invisible God' by whom all things were made (Jn. 1:10), those in heaven and on earth, and in whom all things consist (Col. 1:15 f). [20]

At the Parousia, Winklhofer continues, everything will reflect the mystery of Christ. The cosmos will show with perfect clarity the idea it embodies, the whole message God has written there. And on that day,

> Christ stands forth in all the fullness destined for Him by the Father, inasmuch as He not only takes to Himself all mankind fitted to become His members, but makes of it a single global manifestation of His Spirit and will, so that He reigns through it and it through Him. In this epiphany at the last day, man's dominion over all creation is finally achieved in Christ. [21]

---

[19] Thomas Merton, *The New Man* (New York: Mentor-Omega, 1963), p. 90. We do not know how or to what extent nature will be gathered up in the final Kingdom. Thus Teilhard wrote, "I do not attribute any definitive or absolute value to the varied constructions of nature. What I like about them is not their particular form, but their function, which is to build up mysteriously, first what can be divinized, and then, through the grace of Christ coming down from our endeavor, what is divine," cited by O. Rabut, *Teilhard de Chardin* (New York: Sheed and Ward, 1961), p. 185.

[20] Alois Winklhofer, *The Coming of His Kingdom* (New York: Herder and Herder, 1963), p. 172.

[21] *Ibid.*

Having seen the cosmic significance of these truths of the faith, perhaps we can appreciate better why Teilhard called the universe the "divine milieu" and why he believed so strongly that science and theology must meet in a dialectic.

TOWARD A THEOLOGY OF SCIENCE

It has often been said that even though the Church might have an elevated view of the cosmos, in practice, the Catholic attitude toward human temporal activity is "other-worldly" and destructive of true humanism. There is an element of truth in this allegation. There have always been some in the Church who tend toward a Manichean or Jansenist outlook by regarding human nature and the world with suspicious pessimism. For too long their vivid description of the temptations that accompany human progress has encouraged the belief that the urge toward scientific understanding and mastery of the universe is something basically demonic and unredeemable; that it is the effect of original sin leading man to a final, more disastrous revolt against God. No one denies that progress brings with it temptation, but does it not also provide unlimited potentiality for good? It is perfectly Christian to anticipate risks and dangers, but what a distortion of Catholic theology it is to cower, withdraw and evade responsibility.

The Church's teaching on man's work in the world is a synthesis of two main emphases, each valid and necessary, each capable of being overstressed apart from the other. One, the incarnational view, accents the truth that to be true to the command of Genesis, Christ's Incarnation, and the Church's mission in the world, man must strive to master matter and advance the new creation, so that, aided by the universe in his search for salvation, he may in turn consecrate it back to God. Even activity that is not intentionally Christ-directed can contribute to the coming of the Kingdom. For by enriching the temporal order, by striving for the "hominization" of the world, man prepares the fullness of time that will precede the Second Coming of the Lord. The Christian can never be an escapist; his vocation is to bring the Good News of salvation to the world in the form of redemptive love, and this demands involvement. His task is to Christianize the world from within. Urged on by grace, his human efforts are pledged to guide human progress forward toward Christ and away from Babylon. From the moment Christ ascended to the Father, and the apostles were chided with the question, "Men of Galilee, why stand there looking up into the sky?" (Acts 1:10), the Church's mission has been to bring the grace of Christ to the world and to transpose it with the mark of redeemed man.

On the other hand, there is an eschatalogical emphasis in Catholic theology that prevents man from expecting to build a world that God will only have to bless on the last day. The world as we know it is not man's final dwelling place and he must be careful not to identify so completely with an imperfect world that he will be unable to recognize or share in the perfect world that will be fashioned by Christ. He cannot let himself become so enamored with the trip that he loses sight of the goal. Teilhard remarks that there is nothing so distinctively human in the Christian as his detachment. And this is true because the Christian, while laboring for a more perfect world, is always aware that no purely temporal goal is worthy of his total commitment. He is a pilgrim in the world and until his human nature is totally graced, he will always have to use the weapons of self-denial and discipline in combating the inclination to be led by creation and to lead creation away from God.

By incorporating both the incarnational and eschatalogical attitudes in her integral world-view, the Church encourages a true Christian humanism, the kind described by Teilhard when he wrote:

All I want to say can be summed up in three phrases. Some—the old-fashioned Christians, say: Await the return of Christ. Others—the Marxists, reply: Achieve the world. And the third—the neo-Catholics think: In order that Christ can return, we must achieve the world.[22] In more detail, Yves Congar expresses the task of the Christian in the world:

> The world of men is a world of bitter conflicts and injustices; by engaging in and disengaging from it, immersing himself in and rising above it, in accordance with the law of his kingly priesthood, the Christian has the opportunity to play the part of the peace-maker which the Gospel attributes to God's true children. The world is atrociously divided; progress, bound up with technicized industry, brings an element of division into what should be a uniting factor, and the world itself is characterized by plurality and by rejection of hierarchies. It is a world too of grinding competition, where children soon learn to 'look after Number One.' Hence a frantic utilitarianism which judges only by the biggest and quickest output, which makes it difficult for human character to mature and most elementary ideas to develop.[23]

Yet, Father Congar continues, Christians have a magnificent contribution to make in such a world:

> Provided they are as effectively engaged in it as they are authentically disengaged and disinterested, provided they live in God as truly as they live in the world, it is possible for them to be a factor for peace and unity; their character, their loyalty, their ability, together with their worth as spiritual men, should enable them to be accepted as reliable judges of things, to be a sort of arbitrator. In a frenzied world, it is for them to pursue a long-term policy of true humanism, sanity and serious work, if they believe in God wholeheartedly enough to believe in man as God believes in him, that is, more strongly than any man can do who uses only human standards.[24]

Catholic teaching, it seems to me, demands that we recognize the dignity and ultimate value of the scientific quest. Too few voices have been proclaiming the fact that the Church invites man to believe in himself and in his potentialities. Can we who profess the faith in its fullness fail to acknowledge gratefully the important role science must play in the working out of personal and cosmic redemption? It was Pope Pius XII who said:

[22] Letter to L. Swan, cited in her "Memories and Letters," *Teilhard de Chardin: Pilgrim of the Future*, ed. cit., p. 45, n. 3.
[23] Yves Congar, *Lay People in the Church* (Westminster: Newman, 1956), p. 427.
[24] *Ibid.*

> He [the believer] will even find it natural to place beside the
> gold, frankincense, and myrrh offered by the Magi to the infant
> God, also the modern conquests of technology: machines and
> numbers, laboratories and inventions, power and resources. Fur-
> thermore, such an offer is like presenting Him with the work
> which He Himself once commanded and which is now being ef-
> fected though it has not yet reached its term. "Fill the earth and
> subdue it" (Gen 1:28), said God to man as He handed creation
> over to him in temporary heritage. What a long, hard road from
> then on to the present day, when men can at last say that they
> have in some measure fulfilled the divine command.[25]

Scientific activity as reverence for creation is itself a form of
worship. The scientist is bound to be caught up in the mystery of
creation and, as Mouroux has said, creation is so fair that when
man contemplates it he has no other choice: he must either adore
it or present it to God.[26] A truly scientific understanding of the
universe ultimately requires a religious interpretation, or else one
overly materialistic, that makes man his own creator. Christian
humanism protects the full dignity of man and science while
working to return creation to God. Secular humanism attempts
to exalt man, but to the expense of his dignity as a redeemed
son of God, thus denying his most noble heritage.

The progress of science has put us on the verge of becoming
truly the master of nature instead of its humble dependent. Within
a relatively short space of time we have discovered nuclear energy,
explored the expanses of space, invented electric computers capable of many operations formerly reserved to the human brain.
Advances in biochemistry, psychology, pharmacology, and allied
sciences put us on the threshold of changes almost impossible to
conceive. Teilhard was enthusiastic about the scientific explosion:
"Each day brings me a new evidence that we are playing our part
in the birth of something great. I think that at no other time
has tension for life been so strong." [27] But there is a real question
as to where science is leading us. The alternatives are clear.
Either it will develop along materialistic lines as in Soviet Russia,

[25] Pope Pius XII, Christmas Message, December 24, 1953.
[26] See J. Mouroux, *The Meaning of Man*, ed. cit., p. 44.
[27] Letter cited by L. Swan, *op. cit.*, p. 43.

seeking the absolute conquest of nature and man as the ultimate phase in a terrestrial messianism or, renouncing any claim to univocal validity, it will not only allow other approaches to reality, but will recognize a need to relate to them in a cooperative effort that will be both proximately and ultimately practical for the human race. This is the course that will have to be followed if we are to seek, attain and employ scientific knowledge in a way perfective of the whole man: body and soul, reason and will.

Man's freedom to determine his own destiny is rooted in both the intellect and the will. Knowledge brings freedom from the imprisonment of error, but that is not enough. The will must enter in with its liberty and its principal object, an unselfish love of what is truly good. History dramatizes that brilliant intelligence in the clutches of misdirected will is as disastrous as ignorance misguiding the commitments of good will. Man's integral freedom is toward a more mature human being-in-the-world, toward knowledge rectified by love, and love deepened by knowledge. Intellectual disciplines can perfect the human reason, but only if they originate and terminate in love will they perfect the whole man and serve the aspirations of the human community.

If science cooperates with other human responses to reality and builds the city of man according to the natural perfective norm of reason and love, the resulting progress in the hominization of the world can provide a dynamic disposition for the Christian hope of personal and cosmic salvation that is the fruit of grace.

Because she is committed to bringing Christ's salvific grace to the world and because personal salvation supposes free and fully human acceptance of grace, the Church, living in anticipation of the Parousia, is uniquely able to encourage progress in the entire spectrum of human development. For she sees true human advancement in knowledge and love as a natural disposition for the reestablishment of all things in Christ and for the realization that ultimate personal perfection comes only with the grace of discerning and serving the will of God.

The Church must make her voice heard at this critical moment of history. Through science man holds in his hand the power to decide the future. He must be warned against the possibility of becoming so immersed in matter and material advance that instead of "spiritualizing" matter, he begins to "materialize" what is most human in him, his spirit.[28] More importantly, man has to be encouraged to live up to the dignity of his nature by using his science with moral responsibility to free man instead of chaining him to a new kind of slavery, the servitude of man at the exclusion of God.

In summary, I believe the Church has several contributions to make toward a new spirit of cooperation between religion and science. First, the Church possesses an elevated view of the cosmos that brings out the sacramental nobility of what science studies. Secondly, she teaches a Christian humanism that recognizes the full dignity of science and its role in the return of the universe to God. Thirdly, her mission to bring salvation to the world means that she encourages the use of science for fully human development since this can be a natural disposition for the acceptance of grace.

Teilhard de Chardin challenged both scientists and theologians to realize both the possibility and necessity of effective communication. As a priest-scientist, Teilhard had the uncommon advantage of a double competence that allowed him to unite in the thought-processes of a single mind the data of science and faith. In his thought, science and religion, while remaining autonomous, combined in a dialectic of open-minded creativity that resulted in a brilliant vision of mankind's destiny. We cannot underestimate the importance of that creativity and vision. The future of

---

[28] We must not underestimate the power of the Marxist "theology" of science by reducing it to a complete materialism. Marxism does advocate a spiritual liberty; its aim is to free man from economic and social slavery, leading him to a classless society abounding in equality and justice. But it would also free man from subjection to God and close off that part of his spirit that thirsts for the source of being, goodness, truth and love. Western science does not "prove" God's existence and activity, but, by recognizing the limitations of its method and object, it acknowledges man's freedom to encounter God in faith and love and thus his freedom to accept the gift of redeeming grace.

man may well depend on what we do today to carry on a fruitful encounter of science and religion. We need to develop a middle ground, a theology of science, analogous to the philosophy of science, where clarifications can be made, related theories discussed, and tensions resolved, in an atmosphere of truth-seeking and mutual respect.[29]

Of course, the truths of faith are of another order than those of science and are not subject to scientific proof or disproof any more than scientific truths can be confirmed or denied on the basis of faith. But attempts to explain the faith in human terms and to apply it in daily life as well as efforts to draw out the ultimate significance of scientific knowledge enter an area of common concern where science and religion, together with philosophy, have something to say to each other.

Had there been a theology of science in the seventeenth century, the Galileo case would not have happened. Galileo would not have proceeded without regard for religious hierarchies nor have given the impression that he considered his proof demonstrative. Theologians would have had incentive to appreciate Galileo's kind of science and they might have seen that faith was not being destroyed, but purified by Galileo's discoveries.

There are several ways a theology of science can serve the human community today. Holy Scripture presents Divine Revelation in the cosmological context of the Hebrew "three-decker" universe. This specific but non-essential cosmological framework was merely a vehicle used to convey religious truth to the people.

---

[29] Let me explain my use of the title "theology of science." The term "theology" here does not imply that any of the conclusions of science have been revealed in the Scripture and thus demand direct study by the theologians. The Bible does not teach science. But the truths revealed in Scripture do pertain to science inasmuch as all scientific contributions to mankind are actually made within the existential fact of God's revelation to man of His existence, providence, love and redemption of men through Christ. It is within this context, at least for the believer, that the ultimate value of science can be seen both in itself and in its relation to man. To see science in this way is not to rob it of its hard-won independence from theology, nor an attempt to make science bow before theology. Whether or not scientists who do not profess the faith agree with our conclusions, perhaps they will see in our efforts an honest statement of the basic harmony of science and religion.

It may be that religious truths would be easier for us to understand today if we express them in terms of the universe as we
now know it. The theology of science could undoubtedly help
find the most appropriate analogies for proclaiming the religious
message of God's Revelation in a modern context. This would be
authentic "demythologizing." The fact is that God is speaking to
us through modern secular and scientific civilization, as He has
spoken to us through past cultures. Hence one of the tasks of
the theology of science is to listen to what God is saying to us
through modern science and modern society, and to use this in
understanding the word of God.

Of course, the most obvious contribution a theology of science
can make is in facing the momentous moral problems that con-
front us today, some of them for the first time. The question of
whether or when the hydrogen bomb could be unleashed is not
something politicians can decide by themselves; they need the
professional opinion of theologians who are acquainted with accurate scientific information on such factors as the extent of destruction, the possibility of directing it to a specifically military
target, and its ultimate effect on the human race. Similarly, the
alarming growth in mental illness, the psychological effects of mass
media advertising and some recent trends in literature and cinema,
the population explosion and the use of certain drugs, are but a
few of many problems that demand scientific-theological collabo-
ration for their solutions.

The theology of science offers a common ground where differences of approach and interpretation, tensions in finding the truth,
and the possibility of meta-rational synthesis can be faced hon-
estly and creatively. This is already being done with some success
in several institutes such as the Academy of Religion and Mental
Health, which is founded on a belief in the necessity of an encounter between science and religion.

> We believe that there is a deep and abiding relationship between
> man's faith and man's health and usefulness in the world. We
> believe that the relationship is just beginning to be explored,
> and that it must be explored with teachableness and humility.

We believe that the exploration will prove fruitful for religion and for the medical and behavioral sciences. We believe that the exploration must be conducted in an atmosphere of free inquiry and criticism. And finally we believe that it can be done in a language that thoughtful people can understand.[30]

There are, of course, a number of difficulties that seem to militate against fully developing a theology of science. For one thing, theologians and scientists speak different languages, both demanding precision, and this might block effective communication. Also, both are completely engrossed in their own work and it might seem a luxury to devote time to a new field of writing, lecturing and meetings. In addition, there are so many trends and counter-trends in both theology and science that perhaps nothing definitive could be reached anyway.

No doubt there is some truth in these and similar objections that could be raised. But the question is not so much "Will it be difficult to accomplish?" as it is "Is the venture worth attempting?" Given the crisis of contemporary civilization, the answer to this must be a resounding "Yes."

There is such a scientific emphasis in society today that if the Church is to carry on a dialogue with the modern world, she must take the initiative in promoting communication between religion and science. This must be done not only in Church-wide institutes of a ceremonial character such as the Pontifical Academy of Science,[31] but at the working level of Catholic universities, colleges and theological institutes. Qualified Catholics, religious

[30] H. C. Meserve, "The Encounter between Science and Religion," an editorial in the *Journal of Religion and Health*, vol. IV (October, 1964), p. 5.

[31] The Pontifical Academy of Science consists of seventy members, all eminent scientists, representing many nations and religious beliefs. Once a year, the members gather at the Vatican to discuss current scientific problems and developments. It is interesting to note that the Academy is a direct descendent of the *Accademia dei Lincei*, which was founded by Prince Cesi in 1603, and to which Galileo proudly belonged. In the eighteenth century, the *Accademia* entered a period of decline and, for all practical purposes, disbanded. Seeing the value this organization could have for scientific advancement, Pope Pius IX revived it and sponsored it under the name *Pontificia Accademia dei Nuovi Lincei*. In 1936, Pope Pius XI revitalized it and gave it its present name, the Pontifical Academy of Science.

and lay, should be urged to pursue graduate training in both theology and science.

While the Church has no monopoly on truth, nothing authentically human, whatever its origin, can be foreign to her. Moreover, the Church seems ideally suited for the task of furthering the theology of science, for she possesses both a disciplined structure and the capability of assimilating the insights of countless opposites in a way that fosters unity but does not demand uniformity.

The Second Vatican Council has helped break down the widespread image of the Church as a rigidly dogmatic and authoritarian institution. Many observers have been impressed both by the unity in essentials that manifests a healthy orthodoxy and the honest differences of opinion as to how the Good News of salvation is to be presented and applied in the contemporary world. The Church shows proper concern for the purity of her witness to Divine Revelation, of which, under the guidance of the Spirit, she is the living custodian, interpreter and teacher. In defining the central beliefs of the faith, the Church in no way pretends to exhaust their mystery. When the Council of Trent, for example, defined that the sacraments of the New Law confer grace on those who place no obstacle to its reception, it did not say how this takes place, leaving it to theologians to develop theories of sacramental causality. Thus while all theologians in the Church believe that the sacraments cause grace, some explain this in terms of efficient physical causality, while others use moral or intentional causality in attempting to make the mystery intelligible. Much as the musical structures Bach imposed on himself provided a creative and creating tension within which he found freedom of expression, the authority of the Church guarantees the truth of essential beliefs while inaugurating human investigation into the depth and implications of the divine mysteries. Christ gave doctrinal authority to the Church not to make truth static, but to constitute her as a living rule of faith, thus insuring dynamic witness to the Word of God in the world.

But the very authority that protects the unity and purity of belief can, if overextended, stifle creativity in the community. Because it is such a complex reality, it is difficult to understand the nature of authority in the Church without becoming involved in historical problems connected with its use and abuse. Father John L. McKenzie suggests that we should locate authority and power in the Church within the controlling theme of the New Testament idea of the Church as a community of love.[32] Authority must command in love and subjects obey in love and it is in this context that ecclesiastical rule and Christian obedience are intelligible. Authority is a gift of the Holy Spirit who directs the Church, but it is not His only gift to the Church. "There are different gifts, but the same Spirit; and the one Spirit is the one life of the body, whose members fulfill different functions in harmony. No member can say of the other members that it does not need them." [33] Those in authority need a deep faith in the workings of the Spirit to be open to the truth wherever it is found, and a love that at once allows freedom and protects it.

There is a fresh openness in the Church today, a realization that mature respect for authority is truly the way to freedom, and authority's respect for freedom an indispensable cause of maturity. This gives substance to the hope that the Church will more and more realize her capability of serving as a unifying center in which the intellectual and spiritual advances of mankind may be gathered, related and directed, that they may in turn, and under the guidance of the Spirit, lead the community of man's ascent to God.

As we work together until the moment when all of creation will unite in the spirit of Christ, let us echo the confident prayer of Father Teilhard de Chardin:

> . . . I, your priest, on the altar of the whole earth, offer you, Lord, the toil and sorrow of the world . . . Receive me O Lord, a victim which creation, drawn by your power, offers up to you in today's

[32] See John L. McKenzie, "Authority and Power in the New Testament," *The Catholic Biblical Quarterly*, vol. XXVI (October, 1964), pp. 413 ff.
[33] *Ibid.*, p. 419.

new dawn . . . In the heart of this formless mass of earth, You have planted an irresistible and sanctifying urge which makes each one of us, from the godless man to the man of faith, cry out, 'Lord, make us to be one.' [34]

[34] Teilhard de Chardin, *Hymn of the Universe* in Braybrooke, *op. cit.*, p. 128.

CHRONOLOGICAL
summary of Galileo's life

1564   Galileo born at Pisa, Italy, on February 15.

1575–7 Began formal schooling at the Monastery of Vallombrosa.

1578   Entered the Vallombrosan Order but left before completing the year of novitiate.

1581   Begins studies at the University of Pisa.

1586   Invents a hydrostatic balance.

1588   Writes a treatise on the center of gravity in solids which wins him some acclaim.

1589   With the help of Guidubaldo del Monte, Galileo obtains a professorship of mathematics at the University of Pisa.

1591   Galileo resigns from Pisa after conflicts with Aristotelians. G. del Monte helps him obtain the chair of Mathematics at the University of Padua.

1597   Writes to Kepler that he has been a Copernican "for several years."

1600   Daughter Virginia (later Sister Maria Celeste) born out of wedlock.

1601   Daughter Livia (later Sister Archangela) born.

1605   Returns to Florence during the summer to tutor Prince Cosimo.

1606   Birth of a son, Vincenzio.

1608   Invention of the telescope by Hans Lippershey.

1609   February ....Prince Cosimo becomes Grand Duke of Tuscany.
        July–August ..Constructs a telescope and begins observing the heavens.

1610   March ....publishes the *Sidereus Nuncius*.

June _____resigns from the University of Padua.

September, returns to Florence as Ducal Philosopher and Mathematician to Cosimo II.

1611   Makes triumphant journey to Rome. Jesuit astronomers confirm his discoveries. Wins election to the *Accademia dei Lincei*. Returns to Florence and gets involved in a dispute concerning the behavior of bodies in water.

1612   Publishes discourse on floating bodies and writes letters on the sunspots. Mistakenly accuses Father Lorini of attacking him from the pulpit.

1613   Letters on the sunspots published by the Lincean Academy. Hears from Father Castelli that his doctrine has been challenged on the basis of Holy Scripture at the court of the Grand Duke. Writes Letter to Castelli.

1614   Publicly attacked by Father Caccini.

1615   Letter to Castelli denounced to the Holy Office but judged in favor of Galileo. Father Foscarini publishes a book trying to reconcile the new astronomy with Sacred Scripture. Cardinal Bellarmine writes Letter to Foscarini warning him and Galileo to stay in the area of hypothesis until demonstrative proof is produced. Galileo goes to Rome to defend his position. Thomas Campanella writes his *Apologia pro Galileo* at the request of Cardinal Gaetani.

1616   February 19 __Theological Consultors of the Holy Office summoned to give their opinion on the Copernican doctrine.

February 23 __Consultors censure Copernican opinion as heretical.

February 25 __Pope Paul V assigns Cardinal Bellarmine to tell Galileo not to hold or defend his theory.

February 26 __Date of the famous injunction recorded in the Holy Office files which claims that Galileo was told by the Commissary General not to discuss his theory in any way.

March 3 _____Cardinal Bellarmine gives Galileo a certificate with which to combat the lies which were being spread about him.

March 5 _____Decree of the Congregation of the Index prohibits Copernicus's *De revolutionibus* until corrected and made more hypothetical.

1618   Appearance of the great comets stirs up discussion.

1619   Galileo enters the controversy by writing the "Discourse on Comets" and publishing it under the name of his disciple, Mario Guiducci.

1620   The Congregation of the Index publishes a list of corrections making it possible for anyone to read Copernicus's work.

1621   The deaths of Pope Paul V, Cardinal Bellarmine, and Grand Duke Cosimo II alter the scene considerably. Galileo begins work on *The Assayer* in answer to Father Grassi's *Astronomical Balance*.

1623   Maffeo Cardinal Barberini is elected Pope and takes the name Urban VIII. Galileo dedicates *The Assayer* to him.

1624   Galileo goes to Rome to try to get the Copernican censure revoked. He has six long talks with Pope Urban and is encouraged to write but told to stay within the limits of a hypothetical treatment.

1625   Begins work on the *Dialogue on the Two Great World Systems* which he intends to be "a most ample confirmation" of the Copernican opinion.

1626–   Illness and necessary interruptions prevent him from com-
1629   pleting the *Dialogue*.

1630   January _____Completes the *Dialogue*.

May _____Galileo goes to Rome and works a publishing arrangement with Father Riccardi.

August _____Prince Cesi, Founder of the Lincean Academy and close friend of Galileo, dies.

1631   Galileo sends request to Rome that the printing be done in Florence. Niccolini is able to convince Riccardi to grant the necessary permission.

1632   February _____The *Dialogue* is published.

August _____Sales and publication are halted by order of the Holy Office.

October ....Galileo is summoned to Rome.

1633   February ....Galileo arrives in Rome and is allowed to stay at the Tuscan Embassy.

April ....Questioned twice by Father Firenzuola. Firenzuola and Cardinal Barberini, the Pope's nephew, desire to deal leniently with Galileo.

May ....Galileo gives his defense to the Holy Office. A misleading report on the proceedings is sent to the Pope.

June 16 ....Pope Urban decrees that Galileo is to publicly abjure his opinion and his book is to be prohibited.

June 22 ....Galileo abjures. His sentence was commuted and he was released in the custody of the Archbishop of Siena.

December ......Galileo returns to his Villa at Arcetri, near Florence.

1637   Galileo loses sight in both eyes and has to move into the city of Florence. He continues to work on his new book, the *Two New Sciences*.

1638   The *Discourses on Two New Sciences* is published at Leyden.

1642   January 8 ......Galileo dies.

BIBLIOGRAPHY

PRIMARY SOURCES

Aristotle, *De Caelo*, trans. J. L. Stocks, vol. II of the Oxford translation of Aristotle's works, edited by W. D. Ross, Oxford, 1922.

_____ *Metaphysics*, a revised text with introduction and commentary by W. D. Ross, 2 vols., Oxford, 1924.

Aquinas, Thomas, saint, O.P., *Summa theologiae*, ed. Leonina, *Opera omnia*, IV-XII, Rome, 1888–1906.

_____ *Commentaria in Aristotelis libros de caelo et mundo*, ed. Leonina, vol. III, Rome, 1886.

_____ *Opuscula omnia*, ed. Antonia Redetti, Venice, 1741.

_____ *Sermon on the Creed*, trans. L. Shapcote, *The Three Greatest Prayers*, London: Burns, Oates, 1937.

Augustine, saint, *De actibus cum Felice Manichaeum*, lib. II, PL, XLII, 519–522.

_____ *De genesi ad litteram*, lib. XII, PL XXXIV, 245–486.

Bañez, Domingo, O.P., *Commentaria in secundam secundae Sancti Thomae*, Venice, 1586.

Bellarmine, Robert, saint, S.J., *De controversiis Christianae fidei*, *Opera omnia*, ed. J. Guiliano, Naples, 1856, vol. I.

Bruno, Giordano, O.P., *The Expulsion of the Triumphant Beast*, trans. A. Imeriti, New Brunswick: Rutgers, 1964.

Campanella, Thomas, O.P., *Apologia pro Galileo*, Frankfurt, 1622. This is a most important document relative to the theological aspects of the Galileo case.

_____ *Metaphysics*, Paris, 1638, reissued in phototype by L. Firpo, Turin, 1961.

Caramuel, Johannis, *Theologiae moralis fundamentalis*, vol. I, Leiden, 1676.

Chardin, Pierre Teilhard de, S.J., *The Phenomenon of Man*, trans. B. Wall, New York: Harper and Row, 1959.

―――― *The Divine Milieu*, trans. B. Wall, New York: Harper and Row, 1960.

―――― *The Future of Man*, trans. N. Denny, New York: Harper and Row, 1964.

Clavius, Christopher, S.J., *Opera mathematica*, Mainz, 1611, vol. III.

Copernicus, Nicholas, *On the Revolutions of the Heavenly Spheres*, trans. C. G. Wallis, Great Books of the Western World, Chicago, 1952, vol. XVI, 505–845.

Cremonini, Cesare, *De calido innato*, Biblioth. Vat., St. Louis University List 5:33.

Cordova, Antonio de, *Quaestiones theologicae*, Venice, 1604.

Erasmus, *The Praise of Folly*, trans. J. P. Dolan, *The Essential Erasmus*, New York: Mentor-Omega, 1964.

Galilei, Galileo, *Le Opere di Galileo Galilei*, ed. Nazionale cura et labore A. Favaro, Florence: (1929–1939), 20 vols. This is the classic collection of Galileo's own writings as well as the important documents bearing on his life and work.

―――― *The Assayer*, trans. Stillman Drake, *Discoveries and Opinions of Galileo*, Doubleday, 1957.

―――― *Dialogue on the Great Systems of the World*, trans. G. de Santillana, Chicago University Press, 1953: and the translation of Stillman Drake, Berkeley: the University of California Press, 1962.

―――― *Letter to the Grand Duchess Christina*, trans. Drake, *Discoveries and Opinions of Galileo*.

―――― *Letters on Sunspots*, trans, Drake, *Ibid*.

―――― *The Starry Messenger*, trans. Drake, *Ibid*.

Kepler, John, *Epitome of Copernican Astronomy*, abr. trans. by C. G. Wallis, *Great Books of the Western World*, Chicago, 1952, vol. XVI.

Lapide, Cornelius a, *Commentarium in Josue, Job, Judicum*, Antwerp, 1664.

Leo XIII, Pope, *Providentissimus Deus*, Encyclical Letter (1893), *Enchiridion Biblicum*, 66–119.

Luca, John Baptist de, *Theatrum veritatis et justitiae*, Venice, 1698, vol. XV.

Menochi, Jacobus, *De arbitrariis judicum*, Venice, 1588.

Passerini, Petrus, *Commentarium in quartum de quintum librum Sexti Decretalium*, Rome, 1672, vol. III.

Ptolemy, Claudius, *The Almagest*, trans. R. C. Taliaferro, *Great Books of the Western World*, vol. XVI, 1–478.

SECONDARY SOURCES

Abele, Jean, *Christianity and Science*, New York: Hawthorn, 1961.

Agar, William M., *Catholic and the Progress of Science*, New York, 1940.

Armitage, Angus, *The World of Copernicus*, New York: Mentor, 1961.

Ashley, Benedict, O.P., *Aristotle's Sluggish Earth: The Problematics of the De caelo*, River Forest: Albertus Magnus Lyceum, 1958.

Balthasar, H. U. von, *Science, Religion and Christianity*, London: Burns, Oates, 1958. A thought-provoking book that points out the implications of contemporary science for religious development.

Braybrooke, Neville, (Ed.), *Teilhard de Chardin: Pilgrim of the Future*, New York: Seabury, 1964. Contains several excellent essays on the meaning of Teilhard's life and work.

Brecht, Bertold, *The Life of Galileo*, trans. D. I. Vesey, London, 1960. This play frequently departs from historical accuracy.

Brik, H. T., *Galilei und sein Prozess*, Berlin, 1963.

Brodrick, James, S.J., *Robert Bellarmine, Saint and Scholar*, Westminster: Newman, 1961. A very readable biography containing many insights into the theological milieu of the seventeenth century.

_____ *Galileo*, New York: Harper and Row, 1965. A short, popular account of the man and his misfortunes.

Broglie, L. de, *The Revolution in Physics*, New York: Noonday, 1953. A noted physicist analyzes the impact of the quantum theory on modern physics.

Brophy, J. and Paolucci, H., *The Achievement of Galileo*, New York, 1962.

Burke, R., *What is the Index?* Milwaukee: Bruce, 1952.

Burke-Gaffney, H.N., *Kepler and the Jesuits*, Milwaukee: Bruce, 1944.

Burstyn, Harold, "Galileo's Attempt to Prove that the Earth Moves," *Isis*, June, 1962.

Burtt, Edwin A., *The Metaphysical Foundations of Modern Science*, Garden City: Doubleday, 1955. A brilliant synthesis of the currents of thought that led to and developed with modern science.

Cahill, John, O.P., *The Development of Theological Censures after the Council of Trent* (1563–1709), Fribourg, 1955.

Capello, Felix, *De Curia Romana*, Rome, 1911.

Cartechini, Sixtus, S.J., *De valore notarum theologicarum*, Rome, 1951.

Catholic Commentary on Holy Scripture, New York: Nelson, 1953.

Chauchard, Paul, *Science and Religion*, New York: Hawthorn, 1962.

Cohen, I. Bernard, *The Birth of a New Physics*, Garden City: Doubleday, 1960. A very clear and penetrating introduction to the scientific developments of the seventeenth century.

Congar, Yves, O.P., *Lay People in the Church*, Westminster: Newman, 1956.

Cooper, Lane, *Aristotle, Galileo and the Leaning Tower of Pisa*, Ithaca, Cornell, 1935. A scholarly examination of the evidence for and against the historicity of Galileo's supposed experiment from the Tower of Pisa.

Council Speeches of Vatican II, ed. Y. Congar, H. Kung and D. O'Hanlon, New York: Deus Books, 1964.

Crombie, A. C., *Medieval and Early Modern Science*, Garden City: Doubleday, 1959, two vols. Contains excellent bibliographical material, but on the whole, it is of limited value to beginners.

———— "Galileo's 'Dialogue concerning the Two Principal Systems of the World' " *Dominican Studies*, III, 1950.

_____ ed., *Turning Points in Physics*, New York: Harper Torchbooks, 1961. A collection of lectures by scientists and philosophers of science on major events in the history of physics.

Dahm, M. Charles, O.P., "Sir Isaac Newton and Hypothesis," *Reality* IX, 1962, 172–189.

Davis, Charles, S.J., *Theology for Today*, New York: Sheed and Ward, 1962. A fine presentation of some current developments in Roman Catholic theology.

Defant, Albert, *Ebb and Flow*, Ann Arbor, 1958. A clear explanation of the phenomenon of tides.

Dessauer, F., *Der Fall Galilei und wir*, Frankfurt, 1957.

D'Elia, Pascal, S.J., *Galileo in China*, Cambridge: Harvard U. Press, 1960.

Dijksterhuis, E. J., *The Mechanization of the World Picture*, Oxford, 1961. A valuable study of the changing world view produced by the scientific revolution of the seventeenth century.

_____ and Forbes, R.J., *A History of Science and Technology*, Baltimore: Penguin, 1963, vol. I.

Dillenberger, John, *Protestant Thought and Natural Science*, London: Collins, 1961. Undoubtedly one of the finest works yet published on this subject.

Dingle, Herbert, *The Scientific Adventure*, London: Pitman, 1952.

Donat, Joseph, *The Freedom of Science*, New York: Wagner, 1914. A classic example of the old apologetics.

Dondeyne, Albert, *Faith and the World*, Pittsburgh: Duquesne, 1963. Highly recommended for its precise presentation of what faith is and how it should affect one's view of temporal realities.

Drake, Stillman, *Discoveries and Opinions of Galileo*, Garden City: Doubleday, 1957. A collection of translations from some of Galileo's chief works with introductions by Stillman Drake, the leading Galileo scholar of our time.

_____ "A Kind Word for Sizi," *Isis*, June, 1958.

_____ and Santillana, G. de, "Koestler and His Sleepwalkers," *Isis*, September, 1959.

Dreyer, J. L. E., *A History of Astronomy from Thales to Kepler*, Cambridge, 1919.

Duhem, Pierre, *Essai sur la notion de théorie physique de Platon à Galilée*, Paris, 1908.

————— "To Save the Phenomena," trans. A. Paolucci, *The Achievement of Galileo*, New York: Twayne, 1962.

Einstein, Albert, *Essays in Science*, New York: The Wisdom Library, 1962.

Fahie, J. J., *Galileo, His Life and Work*, London, 1903. Presents a relatively objective picture of Galileo.

Fouille, A., *Historia general de la Filosofia*, Santiago, Chile, 1955.

Frank, Philipp, *Modern Science and its Philosophy*, New York: Braziller, 1955.

Galli, P. M., O.P., "Campanella e Galileo," *Memorie Dominicane*, October, 1953.

Gebler, Karl von, *Galileo and the Roman Curia*, London, 1879. A very scholarly, though not altogether accurate account of the condemnation.

Geymonat, L., *Galileo Galilei*, trans. S. Drake, New York: McGraw-Hill, 1965. A highly knowledgeable interpretation of Galileo's philosophy of science.

Gilson, E., *Christianity and Philosophy*, New York, 1939.

Gordon, W., *Science and Theology*, London, 1933.

Grisar, H., S.J., *Galileistudien*, Ratisbon, 1882. A scholarly but defensive approach to the condemnation.

Hall, A. R., *The Scientific Revolution 1500–1800*, London, 1954. An excellent survey by a respected historian of science.

Heaney, J. J., Ed., *Faith Reason and the Gospels*, Westminster: Newman, 1963. Highly recommended for anyone interested in Christianity and science today.

Haydn, Hiram, *The Counter-Renaissance*, New York: Grove, 1960.

Heisenberg, W., "The Representation of Nature in Contemporary Physics," *Symbolism in Religion and Literature*, ed. Rollo May,

New York: Braziller, 1961. One of a number of articles in this book well worth reading.

Heim, Karl, *Christian Faith and Natural Science*, London: S.C.M., 1953. An important contribution by a noted Protestant theologian.

Henry, A. H., O.P., *Introduction to Theology*, Chicago: Fides, 1954.

Höffding, H., A *History of Modern Philosophy*, trans. B. E. Meyer, New York: Dover, 1955. Thorough and well written.

Horridge, F., *The Lives of Great Italians*, London, 1900.

Hull, Ernest, S.J., *Galileo and His Condemnation*, Bombay, 1913. An attempt to establish precisely where the mistakes were made on both sides.

Jammer, Max, *Concepts of Space*, New York: Harper Torchbooks, 1960.

Journet, Charles, *The Church of the Incarnate Word*, New York: Sheed and Ward, 1954, two vols. A depth study in basic Roman Catholic ecclesiology.

Feynman, R. P., "The Relation of Science and Religion," *Frontiers of Science*, ed. E. Hutchings, New York: Basic Books, 1958. A provocative essay revealing how the author's scientific attitude has affected his religious outlook.

Fichtner, J., *Theological Anthropology*, U. of Notre Dame Press, 1963.

Koestler, Arthur, *The Sleepwalkers*, New York: Macmillan, 1959. A fascinating and well-written book but it is sometimes inaccurate and too highly slanted against Galileo.

Koyré, A., *From the Closed World to the Infinite Universe*, New York: Harper Torchbooks, 1958. A classic study of the development and effects of the new astronomy.

_____ *Études Galiléennes*, Paris, 1939, three vols. Essential for any serious work on Galileo.

Kristeller, P. O., *Renaissance Thought*, New York: Harper and Row, 1961. two vols. A first class evaluation of intellectual movements in the Renaissance period.

Lagrange, M-J., O.P., *Historical Criticism and the Old Testatment*, trans. E. Meyers, London, 1905.

Levie, Jean, S.J., *The Bible, Word of God in Words of Men*, New York: Kenedy, 1961. A fine statement of the Roman Catholic understanding of Holy Scripture as both human and divine.

Lonergan, B. J., *Insight*, New York: Philosophical Library, 1957. A creative work on the theory of knowledge and the philosophical implications of contemporary science.

Lucas, Henry, *The Renaissance and the Reformation*, New York: Harper and Row, second edition, 1963. Provides an excellent historical background for the Age of Galileo.

Mangan, J., *The Life, Character and Influence of Desiderius Erasmus*, New York: Macmillan, 1927, two vols.

Mascall, E. L., *Christian Theology and Natural Science*, New York: Ronald, 1957.

Matson, F. W., *The Broken Image*, New York: Braziller, 1964. Recounts the rise and fall of deterministic science and the effects of the new physics on the general scientific outlook.

McColley, G., "Humanism and Astronomy," in *Toward Modern Science*, ed. R. Palter, New York: Noonday, 1961, vol. II.

McCormick, J., and McInnes, M., *Versions of Censorship*, Garden City: Doubleday, 1962.

McMullin, Ernan, "Galileo and His Biographers," *The Furrow*, XI, 1960, 794 ff.

McSorley, J., *An Outline History of The Church*, St. Louis: Herder, 1943.

Mohan, R., ed., *Technology and Christian Culture*, Washington: Catholic University Press, 1960.

Monsma, J. C., ed., *The Evidence of God in an Expanding Universe*, New York: Putnam, 1958.

\-\-\-\-\-\-\-\-\-\- ed., *Science and Religion*, New York: Putnam, 1962.

Mortier, R., O.P., *Histoire des Maitres Généraux de Frères Prêcheurs*, Paris, 1930, vol. VI.

Mouroux, J., *The Meaning of Man*, Garden City: Doubleday, 1961. A significant book that culls the riches of traditional and contemporary theology with regard to God, man and the universe.

_____ *The Mystery of Time*, New York: Desclee, 1964. A beautiful and profound treatise on a subject of great importance.

Muller, H. J., *Freedom in the Western World*, New York: Harper and Row, 1963.

Newman, John Henry, "Galileo, Scripture and the Educated Mind," *The Essential Newman*, ed. V. F. Blehl, New York: Mentor-Omega, 1963.

Nogar, R. J., *The Wisdom of Evolution*, New York: Doubleday, 1963.

Obler, P., ed., *The New Scientist*, New York: Doubleday, 1962. A collection of essays well worth reading.

Pastor, Ludwig Von, *History of the Popes*, London, 1938, vols. XXVIII and XXIX. Contains many illuminating facts and documents regarding the popes involved in Galileo's conflict with the Church.

Rahner, Karl, S.J., *Free Speech in the Church*, New York: Sheed and Ward, 1959.

_____ *Theological Investigations*, Baltimore: Helicon, 1963, two vols.

Reusch, F. H., *Der Prozess Galilei's und die Jesuiten*, Bonn, 1879.

Riga, P., *Catholic Thought in Crisis*, Milwaukee: Bruce, 1963.

Rosen, Edward, *Three Copernican Treatises*, Columbia, 1939.

_____ "Calvin's Attitude Toward Copernicus," *Isis*, July, 1960.

_____ "Galileo's Misstatements about Copernicus," *Isis*, September, 1958.

Santillana, G. de, *The Crime of Galileo*, U. of Chicago Press, 1955. A very readable but highly slanted version of the condemnation and its causes.

Sarton, George, "The Quest for Truth," *The Renaissance*: A Symposium, New York, 1953.

Schillebeeckx, E., O.P., *Christ the Sacrament of Encounter with God*, New York: Sheed and Ward, 1963.

Schilling, H. K., "A Contemporary Macedonian Plea," *Union Theological Seminary Review*, January, 1963. An intelligent call for cooperation between theology and science.

———— *Science and Religion*, New York: Scribners, 1962.

Schoonenberg, P., *God's World in the Making*, Pittsburgh: Duquesne U. Press, 1964.

Schroeder, H. J., O.P., *Canons and Decrees of the Council of Trent*, St. Louis: Herder, 1941.

Schweibert, E. G., *Luther and His Times*, St. Louis: Concordia, 1950.

Singer, Charles, *A Short History of Science*, Oxford, 1943. A well-regarded book for beginners.

Sullivan, J. W. N., *The Limitations of Science*, New York: Mentor, 1949.

Taylor, H. O., *Thought and Expression in the Sixteenth Century*, New York: Macmillan, 1920, two vols.

Thorndyke, Lynn, *A History of Magic and Experimental Science*, New York: Columbia, 1958, vol. XII.

Walker, D. P., *Spiritual and Demonic Magic from Ficino to Campanella*, London: Wartburg, 1958.

Wallace, W. A., O.P., "Science and Religion," *College Outlines of Sacred Doctrine*, Dubuque, Priory Press, 1962.

Wegg-Prosser, F. R., *Galileo and His Judges*, London, 1889.

Weisheipl, J. A., O.P., *The Development of Physical Theory in the Middle Ages*, New York: Sheed and Ward, 1959.

White, Andrew D., *A History of the Warfare of Science with Theology in Christendom*, New York: Dover, 1960, two vols. A polemical and misleading version of the Galileo affair.

Whitehead, A. N., *Science and the Modern World*, New York: Mentor, 1960. A standard work on the subject.

———— *Symbolism*, New York: Capricorn Books, 1959.

Wohlwill, Emil, *Der Inquisitionsprocess des Galileo Galilei*, Hamburg, 1870.

Index